The Governance of Food Security in Europe

Maria Stella Righettini

The Governance of Food Security in Europe

Collaborative Street-Level Bureaucracies and Learning in Turbulent Times

palgrave
macmillan

Maria Stella Righettini
Political Science, Law and International Studies
University of Padua
Padova, Italy

ISBN 978-3-031-98082-4 ISBN 978-3-031-98083-1 (eBook)
https://doi.org/10.1007/978-3-031-98083-1

This Palgrave Macmillan imprint is published by the registered company Springer Nature Switzerland AG.
The registered company address is: Gewerbestrasse 11, 6330 Cham, Switzerland

Introduction

When the World Health Organization (WHO) declared the COVID-19 pandemic in March 2020, footage of supermarkets being stormed, empty shelves in the shops of Italian cities, and crowds of people frantically stockpiling food left a lasting mark on the author. Those images and people in line to purchase groceries in the subsequent months seemed almost apocalyptic to a generation that grew up during the post-World War II economic boom and that had never experienced the hardships of war and famine. These scenes were a stark reminder of how deeply food is woven into the fabric of human lives—not only as a fundamental resource or an undeniable right but also as a primary necessity that can heavily affect human behavior and redefine relationships within a community. In that moment, it became apparent that food is more than sustenance; food drives both individual actions and societal dynamics, thereby revealing vulnerabilities that are often taken for granted. Prior to the pandemic, successive climate and financial crises compelled public and private actors often for the first time to tackle the social complexities of the food system. However, the COVID-19 pandemic and the rapidly evolving geopolitical and climate landscapes shed a stronger light on the food security issue in Europe, regarding, in particular, the interconnection between public and private dimensions of governance (FAO, 2019). The pandemic crisis triggered urban food emergencies and highlighted the importance of what had been taken for granted before the outbreak of the pandemic. The fragility of food systems and their critical importance for public policies and communities. For countries and cities that rely heavily on complex supply chains to sustain their populations, the disruptions caused by the pandemic exposed weaknesses in food distribution, accessibility, and affordability. Suddenly, policymakers were forced to confront food security issues with renewed urgency as supply bottlenecks, price spikes, and inequality in access to food gave rise to stark divides within communities. Additionally, the emergency showed the need for robust social safety nets, including food assistance programs, subsidies for vulnerable populations, and community-driven solutions to combat food insecurity. However, as the pandemic's aftermath is still being navigated and other ongoing crises are being dealt with, what have Europeans truly learned about building more resilient food systems, and are they better prepared for future shocks in Europe?

Food security exists when *all people, at all times, have physical, social, and economic access to sufficient, safe, and nutritious food that meets their dietary needs and food preferences for an active and healthy life* (FAO, 2019). Food security is a growing global challenge that varies in intensity across countries—including highly industrialized ones. Recurring disruptions due to financial crises, pandemic emergencies, environmental disasters, extreme weather events, and wars have shown that access to food is not only determined by a country's economic strength or the way it allocates resources domestically (Jablonski et al., 2021). While these factors are important, external factors play a critical role in determining food security. The idea of poly-crisis—whereby multiple crises interact with and exacerbate one another—is a powerful analytical tool for understanding the intricate links between food insecurity and turbulent contexts. It highlights the need to acknowledge the multifaceted origins of food insecurity and to investigate the capacity of policy strategies to address systemic interdependencies during and after crises. In Europe, poly-crisis spotlighted the crucial role of urban food systems, prompting policymakers to reconsider and reform multilevel governance frameworks. This shift has underscored the need for integrated policies that address environmental, economic, and social dimensions at national and city levels. However, a substantial knowledge gap remains regarding how multilevel governance structures can be effectively shaped to mitigate food insecurity and shield citizens from external shocks. Bridging this gap requires analyzing the policy implementation processes behind food aid program adoption and change, as well as addressing the challenges posed by the wicked nature of food security—a multifaceted issue, with a high degree of ambiguity regarding goals and values, with multiple causes and solutions, each carrying the risk of unintended consequences (Peters, 2017; Rittel & Webber, 1973). Global measures to curb the spread of COVID-19, such as lockdowns, unexpectedly triggered economic crises, reduced household incomes, and made it harder for individuals and families to access necessities such as food.

The pandemic dramatically underscored the urgency of integrating food issues into a wider European multilevel social policy agenda by magnifying the challenges associated with the instability and precariousness of accessing food in highly industrialized countries. In response, a major escalation in public interventions at multiple governance levels occurred to address the compounded impacts of the pandemic, which had exacerbated existing inequalities and heightened the need for more robust and adaptive food security strategies (Ahn & Norwood, 2020; Cable et al., 2021; Crush & Si, 2020; Dodds et al., 2020). The anti-crisis programs required joint measures involving street-level bureaucrats (SLBs)—such as social workers, teachers, and law enforcement officers—NGO operators, and volunteers to deliver services and benefits to those most in need.

Crises disproportionately impact vulnerable populations, particularly those engaged in precarious work and undeclared work. Food insecurity in advanced economies almost inevitably affects socially marginalized groups, such as the homeless and migrants. The challenge is more serious for low-income households: while not requiring support on a regular basis, they often reduce their food expenditure during crises, signaling deprivation that impacts their quality of life and well-being (ActionAid, 2021). Declining food expenditure is a critical indicator of social hardship: it is a consequence of income insecurity, limited access to affordable food, and the daily struggle to meet basic needs. Multiple health and economic crises have exacerbated vulnerabilities worldwide, thus exposing the fragility of food and social welfare systems. Therefore, public interventions at local, national, and international levels have intensified to mitigate the effects of the pandemic, underlining the need for more flexible and resilient food security strategies (Ahn & Norwood, 2020).

Despite the growing global scientific focus on food crisis and security (Alabi & Ngwenyama, 2023; AlNemer, 2023; Kumareswaran & Jayasinghe, 2022; Swinnen & McDermott, 2020; Zhu et al., 2022), we have barely scratched the surface of how, in turbulent times, food security programs address and adapt to social challenges in a multilevel governance setting (Rabbi et al., 2023; Smith & Thompson, 2023). The author argues that the role of public institutions and frontline workers in policymaking and service provision during crises, particularly in addressing food security challenges in Europe, has received insufficient attention. Analyzing how street-level bureaucrats respond to crises, exercise discretion, and engage in collaborative decision-making with beneficiaries provides fresh and valuable insights into the governance dynamics and food security program design. Building on existing literature, which remains fragmented and often fails to capture the dynamics of food security crises, this book explores the adaptive processes and strategies used to create resilient and responsive governance systems for addressing increasingly fast, complex, and evolving needs. Despite growing social imbalances, food security remains a peripheral policy concern rather than a central priority in contemporary democratic governance. It ought instead to be seen as a priority to tackle growing turbulence and multiple interdependent crises (AlNemer, 2023; Béné, 2020; Cable et al., 2021; López Cifuentes & Fiala, 2022). The lack of a cohesive and robust policy governance framework has resulted in missed opportunities to empirically analyze processes through which public actors can prioritize food security, activate street-level programs, and protect fundamental human rights. Too often has the literature overlooked critical questions such as these: how are public food security strategies and programs conceived, designed, adopted, and implemented on the ground during crises? What does the street-level management during crises teach us about the capacity to learn and design more effective food security programs? Our governance framework is built on a shared and well-established definition of food security governance as a complex, multidimensional concept (Candel, 2014, p. 598). This foundation enables us to critically examine the evolving dynamics that shape, transform, and differentiate governance during crises. By integrating perspectives from street-level bureaucracy, learning theories, and consulting collaborative governance literature, this book explores how governance mechanisms adapt in response to external shocks and turbulences. The book provides a robust analytical framework to assess whether and how food security programs evolved in their design and implementation during crises, considering the interconnections within the European multilevel governance system. In doing so, it highlights the conditions that support stable changes in interactions among the public, key actors, and beneficiaries, thus ultimately shaping food security strategies. The empirical research investigates how crises can foster or hinder transformative mechanisms in food security governance at European, national, and municipal levels. The book examines the crucial connection between systemic crises contexts, their challenges to food security program design, and the grassroots mechanisms at the different governance levels that turn new public intervention theories into practical solutions. It highlights how the decision to collaborate in tackling food insecurity, in turbulent and uncertain times, aims to guarantee the well-being of entire communities and meet the need to increase trust and learning.

Before the pandemic, food security research was dynamic and diverse (Righettini & Bordin, 2023): it offered a multifaceted understanding of the issue from both local and global perspectives. On the one hand, studies focused on micro-level domestic programs aimed at improving diet quality at the household level (Fanzo & Davis, 2021; Leyna et al., 2008; Wrieden et al., 2007). On the other hand, researchers explored the links between micro-local and macro-global dynamics of agricultural systems, such as globalized food supply chains, the impact of biotechnology innovations, and resilience to climate change (McKeon, 2015; Alabi & Ngwenyama, 2023). While earlier studies often focused on either household nutrition or global food systems, recent research has increasingly sought to bridge these macro and micro perspectives to develop more comprehensive solutions to food insecurity. As the 'meso' level of analysis—the policy domain—gains prominence, scholars have traditionally focused on food security in rural areas, examining the challenges faced by vulnerable rural populations in developing countries and the government strategies to support farmers during production crises. The growing interest in urban transitions and their food security dynamics offers a compelling framework for exploring rural-urban interactions and beyond. In fact, with their dense populations and intricate networks, cities serve as crucial arenas for understanding how global trends intersect with local policymaking (Prosperi et al., 2015). Therefore, urban food security programs during crises provide a unique lens for analyzing the interaction between urban street-level bureaucrats, stakeholders, and beneficiaries (Ben Hassen & El Bilali, 2022; Kovács et al., 2022; Rabbi et al., 2023).

The book explores how collaboration across different levels of governance has become a crucial resilience mechanism, transforming these workers into key actors who rapidly adapt to mitigate crises and keep social support systems functioning (Cox et al., 2021; Gofen & Lotta, 2021). It offers a fresh perspective on how food security governance has adapted to the challenges of a poly-crisis context, examining how policies are shaped at street level across various levels of government in response to crises, from the European to the urban scale. It highlights the pivotal role of 'street-level' officials, whose direct interactions with communities influence the development and implementation of food security policies. By examining the complexity of multilevel governance, the book highlights the often overlooked role of frontline workers in adapting policy strategies and tools in response to crises (Lotta et al., 2023; Maynard-Moody & Musheno, 2000). By doing so, the book sheds light on how food security governance and management evolve in addressing routine disruptions and facing emerging real-world challenges (Bennett & Howlett, 1992; Gerlak & Heikkila, 2011; Rose, 1991). The research takes a mechanistic approach, moving beyond traditional focuses, such as cultural or organizational characteristics, professional roles in bureaucracies (Davidovitz et al., 2021; Gofen & Lotta, 2021; Gofen et al., 2019), 'discretion' in public service delivery (Bertelli, 2021; Lotta et al., 2023), or the influence of politicians' and bureaucrats' ideologies on policy implementation (Bertelli, 2006; Stensöta, 2011). Specifically, it focuses on SLBs' interactions with program beneficiaries—both intermediate and final—in the management of Food Aid Programs (Brodkin, 2012, 2021; Evans & Hupe, 2020). By addressing a gap in the literature on street-level bureaucrats, the book offers an original contribution on how crises trigger changes in content, tools, and relationships within food security governance. It does so by focusing on the role of street-level operators in fostering learning and shaping collaborative initiatives (Hupe & Hill, 2007, 2015, 2019; Hupe & Buffat, 2014; Moyson et al., 2017).

The Structure of the Work

Chapter 1 lays the groundwork for the study by defining key concepts such as food security, governance, street-level bureaucrats, trust, and collaborative learning, while introducing the analytical framework guiding the research. It presents the central research question and outlines the methodology, highlighting the integration of multiple approaches to explore learning processes at the street level within food security programs. The mechanistic approach is used to trace the relationships between the poly-crisis, street-level bureaucrats managing food aid programs, and the learning process based on interactions with program beneficiaries.

Chapter 2 examines the evolution of European food security governance before, during, and after the COVID-19 pandemic. It explores the EU's tools and mechanisms, focusing on the transition and management of the Fund for European Aid to the Most Deprived (FEAD), shifting toward a socially integrated approach to addressing food vulnerability. The chapter provides a detailed analysis of 125 FEAD-funded best practices implemented between 2014 and 2021, highlighting how the FEAD multilevel network was constructed and adapted in response to crises across Europe to foster multilevel learning. Moreover, the content analysis of best practices served to deepen our understanding of the European

Commission's approach to food security. It reveals the diverse and adaptive strategies used by local actors to tackle food insecurity, emphasizing the importance of collaboration among public, private, and civil society organizations.

Chapter 3 focuses on Italy as a national case study, examining how the pandemic reshaped food security policy governance. The chapter explores how the emergency prompted multilevel interventions, leading all Italian municipalities to adopt Food Aid Programs (FAPs). Drawing on real-time evidence collected during the two pandemic waves in 2021, as well as in-depth interviews with local government officials, professionals, and partners from the private sector, the chapter highlights how Italian provincial capital cities implemented the national strategy at the local level. It also delves into how local municipalities adapted the food stamp system to meet the surge in demand during the pandemic. It also identifies meso-level mechanisms that emerge within the collaborative learning experience at the local level, which is particularly apparent in the way local government officials, local social services, and partners from the private sector worked together to hone the delivery of food stamps.

Chapter 4 presents two contrasting urban case studies of Milan and Warsaw, examining how each city navigated crises. Using the comparative approach of two opposite cases, it contrasts the two cities' food security narratives—Milan's integrated urban food policy system versus Warsaw's city silos social welfare system—and explores their implications for governance in a poly-crisis context. Despite the major similarities in the emergency intervention tools used by both cities, the chapter emphasizes the differences in the learning mechanisms that emerged at the urban level. These distinctions reveal how the legacy in each city's governance model shaped their response to food insecurity during crises. A key aspect of this comparison is also the influence of centralized versus decentralized governance structures on learning mechanisms and collaboration at the street level. Overall, the chapter notes that the centralized or decentralized nature of governance systems shapes the way street-level bureaucrats and other local actors learn and collaborate in response to food security challenges. It highlights the importance of flexible, collaborative learning mechanisms, which are more likely to emerge in decentralized systems. It also points out the challenges posed by more centralized, top-down approaches to governance.

Chapter 5 includes final considerations on the insights drawn from the evidence collected at European, national, and urban levels. By examining how collaborative learning mechanisms function during crises within governance at the different levels, the chapter highlights both similarities and differences in governance approach. It also offers recommendations for policymakers on how to enhance adaptive food aid programs in turbulent times, thereby ultimately improving their responsiveness and effectiveness. A key element discussed in this chapter is the critical role of poly-crises in reshaping food security governance and the importance of multilevel collaborative learning in enhancing the effectiveness and adaptability of food aid programs.

References

Action Aid. (2021). *La fame non raccontata*. Action Aid, Milano.

Ahn, S., & Norwood, F. B. (2020). Measuring food insecurity during the COVID-19 pandemic of Spring 2020. *Applied Economic Perspectives and Policy, 43*, 162–168.

Alabi, M. O., & Ngwenyama, O. (2023). Food security and disruptions of the global food supply chains during COVID-19: Building smarter food supply chains for post COVID-19 era. *British Food Journal, 125*(1), 167–85.

AlNemer, H. A. (2023). The COVID-19 pandemic and global food security: A bibliometric analysis and future research direction. *International Journal of Social Economics, 50*(5), 709–724.

Béné, C. (2020). Resilience of local food systems and links to food security–A review of some important concepts in the context of COVID-19 and other shocks. *Food Security, 12*, 1–18.

Ben Hassen, T., & El Bilali, H. (2022). Impacts of the Russia-Ukraine war on global food security: Towards more sustainable and resilient food systems? *Foods, 11*(15), 2301. Retrieved December 2, 2024, from https://www.mdpi.com/2304-8158/11/15/2301.

Bennett, C. J., & Howlett, M. (1992). The lessons of learning: Reconciling theories of policy learning and policy change. *Policy Sciences, 25*(3), 275–94.

Bertelli, A. (2006). The role of political ideology in the structural design of new governance agencies. *Public Administration Review, 66*(4), 583–95.

Bertelli, A. M. (2021). *Democracy administered: How public administration shapes representative government*. Cambridge University Press.

Brodkin, E. Z. (2012). *Reflections on street-level bureaucracy: Past, present, and future.* Wiley Online Library.

Brodkin, E. Z. (2021). Street-level organizations at the front lines of crises. *Journal of Comparative Policy Analysis: Research and Practice, 23*(1), 16–29. https://doi.org/10.1080/13876988.2020.1848352.

Cable, J., Jaykus, L. A., Hoelzer, K., Newton, J., & Torero, M. (2021). The impact of COVID-19 on food systems, safety, and security—a symposium report. *Annals of the New York Academy of Sciences, 1484*(1), 3–8.

Candel, J. J. (2014). Food security governance: A systematic literature review. *Food Security, 6*, 585–601.

Cox, R., Dickson, D., & Marier, P. (2021). Resistance, innovation, and improvisation: Comparing the responses of nursing home workers to the COVID-19 pandemic in Canada and the United States. *Journal of Comparative Policy Analysis, 23*(1), 41–50.

Crush, J., & Si, Z. (2020). COVID-19 containment and food security in the global South. *Journal of Agriculture, Food Systems, and Community Development, 9*(4), 1–3.

Davidovitz, M., Cohen, N., & Gofen, A. (2021). Governmental response to crises and its implications for street-level implementation: Policy ambiguity, risk, and discretion during the COVID-19 pandemic. *Journal of Comparative Policy Analysis: Research and Practice, 23*(1), 120–30.

Dodds, K., Broto, V. C., Detterbeck, K., Jones, M., Mamadouh, V., Ramutsindela, M., Varsanyi, M., Wachsmuth, D., & Woon, C. Y. (2020). *The COVID-19 pandemic: Territorial, political and governance dimensions of the crisis.* Taylor & Francis.

Evans, T., & Hupe, P. L. (2020). *Discretion and the quest for controlled freedom.* Springer.

Fanzo, J., & Davis, C. (2021). *Global food systems, diets, and nutrition: Linking science, economics, and policy* (1st edn). 2021 edizione. Palgrave Macmillan.

FAO. (2019). *FAO framework for the urban food agenda.* FAO Knowledge Repository website. Retrieved November 17, 2024, from https://openknowledge.fao.org/handle/20.500.14283/ca3151en.

Gerlak, A. K., & Heikkila, T. (2011). Building a theory of learning in collaboratives: Evidence from the everglades restoration program. *Journal of Public Administration Research and Theory, 21*(4), 619–44.

Gofen, A., & Lotta, G. (2021). Street-level bureaucrats at the forefront of pandemic response: A comparative perspective. *Journal of Comparative*

Policy Analysis: Research and Practice, 23(1), 3–15. https://doi.org/10.1080/13876988.2020.1861421.

Gofen, A., Sella, S., & Gassner, D. (2019). Levels of analysis in street-level bureaucracy research. In *Research handbook on street-level bureaucracy.* Edward Elgar Publishing.

Hupe, P., & Buffat, A. (2014). A public service gap: Capturing contexts in a comparative approach of street-level bureaucracy. *Public Management Review, 16*(4), 548–569.

Hupe, P., & Hill, M. (2007). Street-level bureaucracy and public accountability. *Public Administration, 85*(2), 279–299.

Hupe, P., & Hill, M. (2015). *Understanding street-level bureaucracy.* Policy Press.

Hupe, P., & Hill, M. (2019). Positioning street-level bureaucracy research, In *Research handbook on street-level bureaucracy* (pp. 15–30). Edward Elgar Publishing.

Jablonski, B. B. R., Casnovsky, J., Clark, J. K., Cleary, R., Feingold, B., Freedman, D., Gray, S., Romeiko, X., Olabisi, L. S., Torres, M., van den Berg, A. E., Walsh, C., & Wentworth, C. (2021). Emergency food provision for children and families during the COVID-19 pandemic: Examples from five US cities. *Applied Economic Perspectives and Policy, 43*(1), 169–84.

Kovács, E. K., Bachórz, A., Bunzl, N., Mincyte, D., Parasecoli, F., Piras, S., & Varga, M. (2022). The war in Ukraine and food security in Eastern Europe. *Gastronomica: The Journal of Food and Culture, 22*(3), 1–7. Retrieved December 2, 2024, from https://online.ucpress.edu/gastronomica/article-abstract/22/3/1/192112.

Kumareswaran, K., & Jayasinghe, G. Y. (2022). Systematic review on ensuring the global food security and Covid-19 pandemic resilient food systems: Towards accomplishing sustainable development goals targets. *Discover Sustainability, 3*(1), 29. https://doi.org/10.1007/s43621-022-00096-5.

Leyna, G. H., Mmbaga, E. J., Mnyika, K. S., & Klepp, K. I. (2008). Validation of the Radimer/Cornell food insecurity measure in rural Kilimanjaro, Tanzania. *Public Health Nutrition, 11*(7), 684–89. https://doi.org/10.1017/S1368980007001267.

López Cifuentes, M., & Fiala, V. (2022). Covid-19 as a chance for more food democracy in European Cities? The responses of actors within Vienna's urban food system to the pandemic. London: Cities, *131*, 104041. https://doi.org/10.1016/j.cities.2022.104041.

Lotta, G., Maynard-Moody, S., & Musheno, M. (2023). Practicing knowledge-agency in highly vulnerable communities: The frontlines of governance in delivering Brazilian human services. *Public Management Review*, 1–21. https://doi.org/10.1080/14719037.2023.2288250.

Maynard-Moody, S., & Musheno, M. (2000). State agent or citizen agent: Two narratives of discretion. *Journal of Public Administration Research and Theory, 10*(2), 329–58, https://doi.org/10.1093/oxfordjournals.jpart.a024272.

McKeon, N. (2015). *Food Security governance. Empowering communities, regulating corporations.* Routledge.

Moyson, S., Scholten, P., & Weible, C. M. (2017). Policy learning and policy change: Theorizing their relations from different perspectives. *Policy and Society, 36*(2), 161–77. Retrieved December 2, 2024, from https://academic.oup.com/policyandsociety/article-abstract/36/2/161/6403937.

Peters, B. G. (2017). What is so wicked about wicked problems? A conceptual analysis and a research program. *Policy and Society, 36*(3), 385–96. Retrieved December 2, 2024, from https://academic.oup.com/policyandsociety/article-abstract/36/3/385/6407931.

Prosperi, P., Moragues-Faus A., Sonnino, R., & Devereux, C. (2015). Measuring progress towards sustainable food cities: Sustainability and food security indicators. Report of the ESRC financed Project "Enhancing the Impact of Sustainable Urban Food Strategies.

Rabbi, M. F., Ben Hassen, T., El Bilali, H., Raheem, D., & Raposo, A. (2023). Food security challenges in Europe in the context of the Prolonged Russian–Ukrainian Conflict. *Sustainability, 15*(6), 4745.

Righettini, M. S., & Bordin, E. (2023). Exploring food security as a multidimensional topic: Twenty years of scientific publications and recent developments. *Quality & Quantity, 57*(3), 2739–2758.

Rittel, H. W. J., & Webber, M. M. (1973). Dilemmas in a general theory of planning. *Policy Sciences, 4*(2), 155–69. https://doi.org/10.1007/BF01405730.

Rose, R. (1991). What is lesson-drawing? *Journal of Public Policy, 11*(1), 3–30. Retrieved May 28, 2021, from https://www.jstor.org/stable/4007336.

Smith, D., & Thompson, C. (2023). The complexities and challenges of researching hunger in wealthier countries. In *Food deserts and food insecurity in the UK* (pp. 1–7). Routledge.

Stensöta, H. O. (2011). Political influence on street-level bureaucratic outcome: Testing the interaction between bureaucratic ideology and local community political orientation. *Journal of Public Administration Research and Theory, 22*(3), 553–71.

Swinnen, J., & McDermott, J. (2020). Covid-19 and global food security. *EuroChoices*. https://doi.org/10.1111/1746-692X.12288

Wrieden, W. L., Anderson, A. S., Longbottom, P. J., Valentine, K., Stead, M., Caraher, M., Lang, T., Gray, B., & Dowler, E. (2007). The impact of a community-based food skills intervention on cooking confidence, food preparation methods and dietary choices–an exploratory trial. *Public Health Nutrition, 10*(2), 203–11.

Zhu, X., Yuan, X., Zhang, Y., Liu, H., Wang, J., & Sun, B. (2022). The global concern of food security during the COVID-19 pandemic: Impacts and perspectives on food security. *Food Chemistry, 370*, 130830. Retrieved May 27, 2024, from https://www.sciencedirect.com/science/article/pii/S0308814621018367

Contents

1	**Collaborative Learning from the Street Level: Multi-actor Trust and Collaboration in Uncertain Times**	1
	1.1 Understanding Food Security Governance: A Complex Multidimensional Concept	2
	1.2 Collaborative Governance and Food Security Crises: Opportunities and Challenges	8
	1.3 Street-Level Bureaucrats: From Routine Operations to Crisis Response	11
	1.4 Bridging the Gap: Policy Learning and Street-Level Practices	16
	1.5 Learning from the Unexpected: Collaborating with 'Non-Experts'	19
	1.6 Connecting the Dots: Mechanisms Driving Collaborative Learning	24
	References	39
2	**Food Insecurity in Europe: Governing Amid a Poly-crisis**	55
	2.1 Shifting Focus: Rising Awareness of Food Vulnerability	56
	2.2 From Agriculture to Social Governance Salience: The Evolution of EU Food Aid Governance	61
	2.3 The FEAD Network: Strengthening Social Cohesion in Times of Crisis	67

2.4 Closing Local Gaps: Street-Level Implementation of FEAD Programs 70
2.5 Actors on the Frontlines: Who Shapes FEAD Programs 78
2.6 Beneficiaries' Voices: Navigating the Food Aid System 85
2.7 Mastering FEAD Governance: Proven Mechanisms for Effective Implementation 92
References 97

3 Lessons from Managing the Food Aid Program During the Pandemic: Insights from Italy 103
3.1 Crisis Response in Action: Street-Level Challenges in Managing Food Aid 107
3.2 The National Food Aid Program: Design and Framework 110
3.3 FAP Knowledge Management from the Street Level: Weaknesses and Strengths 114
3.4 Adapting to Change: The Evolution of Municipal Take-Up Processes 115
3.5 Streamlining Administration During the Pandemic: Enduring Reforms or a Return to Routine? 124
3.6 From Food Aid to Food Income: A Shift in Approach 126
References 129

4 The Pandemic as a Game Changer in Urban Food Security Policies 133
4.1 Urban Food Security Governance: A Cornerstone of Municipal Social Equity and Welfare 135
4.2 A New Knowledge Paradigm: Urban Food Security Governance in Crises 139
4.3 Poland's Challenges in a Poly-crisis Era 144
4.4 The Mechanisms of Food Assistance in Warsaw 150
4.5 Milan's Urban Food Security Program: Rising to Turbulent Times 161
4.6 What the Case Studies Reveal 169
References 171

5 Conclusions: Solving the Puzzle of Street-Level Learning During Crisis 177
5.1 Street-level Learning Under Pressure 178
References 189

References 191

Index 195

List of Figures

Fig. 2.1 Global hunger trend from 2005 to 2022. Pre- and post-COVID-19 levels. (Source: FAO, 2023, p. 8, https://openknowledge.fao.org/server/api/core/bitstreams/f1ee0c49-04e7-43df-9b83-6820f4f37ca9/content/state-food-security-and-nutrition-2023/food-security-nutrition-indicators.html) 56

Fig. 2.2 Countries' total available FEAD resources for the years 2016–2021 (mm) and the total number of projects carried out in the same years. (Source: Cohesion Open Data Platform, 2021, https://cohesiondata.ec.europa.eu/stories/s/In-profile-Fund-for-European-Aid-to-the-Most-Depri/tdry-xg55/ and European Union FEAD case studies 2016, 2017, 2018, 2019, 2021) 72

Fig. 2.3 The AROPE rate trends (2014–2020) for the five European countries receiving the most financial support from the FEAD. (Source: Made by the author based on Eurostat (2024), https://ec.europa.eu/eurostat/statistics-explained/index.php?title=Living_conditions_in_Europe_-_poverty_and_social_exclusion# Source_data_for_tables_and_graphs) 73

Fig. 2.4 Interlinks between topics in FEAD projects. (Source: Made by the author based on European Union FEAD case studies (2016, 2017, 2018, 2019, 2021a, 2021b)) 76

Fig. 2.5 FEAD projects based on the relevance of food security measures (2016–2021). (Source: Made by the author based on European Union FEAD case studies (2016, 2017, 2018, 2019, 2021a, 2021b)) 77

Fig. 2.6 Number of street-level bureaucrats identified in FEAD case studies' projects divided by country and by typology. (Source: Made by the author based on European Union FEAD case studies (2016, 2017, 2018, 2019, 2021a, 2021b)) 81
Fig. 2.7 Total number of street-level bureaucrats identified in FEAD case studies' projects divided by the type of measures. (Source: Made by the author based on European Union FEAD case studies (2016, 2017, 2018, 2019, 2021a, 2021b)) 82
Fig. 2.8 Number of FEAD case studies' projects divided by country and the type of social policy field. (Source: Made by the author based on European Union FEAD case studies (2016, 2017, 2018, 2019, 2021a, 2021b)) 82
Fig. 2.9 FEAD case studies' projects divided by country and by the governance level that plans and oversees each project. (Source: Made by the author based on European Union FEAD case studies (2016, 2017, 2018, 2019, 2021a, 2021b)) 83
Fig. 2.10 Target group categories of FEAD projects. (Source: Made by the author based on European Union FEAD case studies (2016, 2017, 2018, 2019, 2021a, 2021b)) 85
Fig. 2.11 Composition of target groups by country. (Source: Made by the author based on European Union FEAD case studies (2016, 2017, 2018, 2019, 2021a, 2021b)) 86
Fig. 2.12 Target groups by year. (Source: Made by the author based on European Union FEAD case studies (2016, 2017, 2018, 2019, 2021a, 2021b)) 88
Fig. 2.13 Total number of stakeholders and street-level bureaucrats mentioned in FEAD case studies' projects divided by country. (Source: Made by the author based on European Union FEAD case studies (2016, 2017, 2018, 2019, 2021a, 2021b)) 90
Fig. 3.1 The multilevel governance framework of the Italian national food aid program during COVID-19. (Source: Made by the author based on official documents) 112

List of Tables

Table 1.1	Food security governance: substantive and procedural dimensions	5
Table 1.2	Concatenation of mechanisms (CM) for building collaborative learning and trust from the street level in the food aid program	26
Table 2.1	EU food security governance transition: structure and mechanisms. From the PEAD to FEAD and ESF+ (1987–2027)	64
Table 3.1	Learning challenges identified by the Italian municipal officials and operators during the first COVID-19 wave and FAP management	116
Table 3.2	Net change in beneficiary eligibility criteria between the first and second food voucher delivery	118
Table 3.3	Changes in food voucher formats between the first and second pandemic waves (99 cities, April 2020 to April 2021)	120

CHAPTER 1

Collaborative Learning from the Street Level: Multi-actor Trust and Collaboration in Uncertain Times

In recent decades, food security has become increasingly significant in policy studies and governmental action, as it is closely linked to public health challenges arising from economic and social crises, both prior to and during the COVID-19 pandemic (Ahn & Norwood, 2020; Cable et al., 2021; Mishra & Rampal, 2020; Moseley & Battersby, 2020; O'Hara & Toussaint, 2021).

Unlike the close-knit network of experts, businesses, and regulatory agencies falling under food safety governance, food security governance operates in a more fragmented, dispersed, rights-protection-driven framework. On the one hand, it adopts a global, macro-level perspective, examining the role of multinational corporations and international organizations such as the Food and Agriculture Organization (FAO) and the WTO in shaping globalized food policies, which have a negative impact on food sovereignty, smallholder farming and food security in rural developing countries (Windfuhr & Jonsén, 2005; Windfuhr, 2010; McKeon, 2015). On the other hand, it embraces a fragmented, micro-level focus, analyzing food poverty at household and local levels by considering the role of local charities, farmers, religious groups, third-sector institutions, and grassroots volunteers, all of which contribute to food security efforts (Carlson et al., 1999; Radimer et al., 1992; Radimer, 2002). Moreover, historically, food governance scholars have focused on the prominent role of the third sector, social movements, and volunteer efforts in addressing

M. S. Righettini, *The Governance of Food Security in Europe*,
https://doi.org/10.1007/978-3-031-98083-1_1

issues related to food security (McKeon, 2015). The term *glocal* encapsulates the antinomy of macro-micro approaches, which often fails to address a crucial yet overlooked aspect: the role of ***institutional and public policy*** in defining and addressing food insecurity in urban interventions.

As a result, a comprehensive analytical framework linking localized findings to broader public policy mechanisms is still missing. The urban governance gap in food security has been identified as a critical, multidimensional challenge. Addressing this gap requires aligning policy sectors and responsibilities, promoting multi-stakeholder participation, developing an integrated communication and information system, as well as mapping and assessing the impact of local initiatives (Candel et al., 2016).

This chapter outlines the conceptual and methodological framework for analyzing the recent evolution of food security governance, focusing on the critical role street-level bureaucrats (SLBs) play in fostering learning and building trust during crises. It puts forward a policy-oriented approach to food security governance where street-level collaboration and trust act as driving forces during crises, while creating a foundation for learning from experience and continuous program improvement. This work's approach draws on multiple bodies of literature: particularly on collaborative governance (CG) (Ansell et al., 2017, 2020; Ansell, 2012; Ansell & Gash, 2008), and SLBs as citizen-agents and key facilitators of collaborative learning (CL) and trust-building (Maynard-Moody & Musheno, 2000, 2022; Meza et al., 2021). SLBs are expected to operate on the frontlines of policy delivery. Their ability to foster collaboration becomes even more critical during a crisis, as they ensure that lessons are learned, trust is maintained, and food security initiatives remain effective and adaptable to emerging challenges (Davidovitz & Cohen, 2022a, 2022b; Cairney & Wellstead, 2021; Klijn et al., 2010).

1.1 Understanding Food Security Governance: A Complex Multidimensional Concept

Food security is a multifaceted and multidimensional issue. Depending on the dimension considered, different actors and potentially different interventions and policies emerge and are relevant for institutional learning. The most used and shared definition of food security was developed during the 1996 World Food Summit organized by the Food and Agriculture Organization (FAO) (Mechlem, 2004). Food security is achieved when 'all people, at all times, have physical and economic access to sufficient, safe, and nutritious food to meet their dietary needs and food preferences for an active and healthy life' (FAO, 2006). The concept has five conceptual dimensions: *nutritional status*, *utilization*, *accessibility*, *availability*, and *stability* (Gross et al., 2000) thus defined: *nutritional status* in the condition of an individual's health based on nutrient intake and utilization, including factors such as food consumption adequacy, nutrient absorption, and the balance between intake and needs; *availability* is the presence of sufficient quantities of quality food, provided through domestic production, imports, or food aid; *access* is the ability of individuals to obtain adequate resources (entitlements) to acquire a nutritious diet, influenced by legal, political, economic, and social factors; *utilization* is the effective use of food through a balanced diet, clean water, sanitation, and healthcare to achieve nutritional well-being and meet all physiological needs; *stability* is the consistent availability and access to adequate food, without the risk of sudden shocks (e.g., economic, climatic crises) or cyclical events (e.g., seasonal food insecurity).

Compared to food insecurity, the concept of 'food poverty' has a descriptive meaning and an economic origin. It indicates food purchasing power, the condition of an individual or household whose income is below a certain threshold and therefore cannot meet the minimum essential needs of the individual or household (O'Connor et al., 2016). Conversely, 'food insecurity' may be defined as a relational concept whose aim is to identify dynamics and interconnections between the factual state of poverty (of individuals and households) and the policy tools to guarantee access to fundamental resources whose purpose is to safeguard and entail human rights.

Food insecurity is a multifaceted challenge that brings together two critical concerns: the basic human right to access food, particularly for those who live in extreme poverty, and the health and well-being of individuals who, while not financially constrained, live in 'food deserts' or have low-quality, nutritionally poor diets (Smith & Thompson, 2023). On the one hand, it emphasizes the urgency of providing vulnerable populations with consistent access to nutritious food. On the other, it underscores the broader public health challenge of improving dietary quality for individuals who can afford food but remain at risk owing to the low nutritional value of their consumption patterns. Its dual nature makes food insecurity a major problem that requires comprehensive solutions that address economic, behavioral, and health dimensions. The various food insecurities identified in the literature by international bodies engaged in the matter and by national and local governments reflect differences in meaning and formulation of the problem and come from different geographical and disciplinary perspectives (Dowler & O'Connor, 2012).

One key issue that warrants attention is the limited studies conducted on food security governance at local and institutional levels before the pandemic (Candel, 2014). Traditionally, the concept was largely associated with mechanisms facilitating debate, convergence, and coordination at global and European levels, as well as frameworks for good governance in humanitarian assistance. It also encompassed the formal and informal rules governing decision-making related to nutritional quality. Pre-pandemic food governance included recurring themes and institutional dynamics, including coordination, convergence, deliberation, the integration of democratic values, and the management of a complex multiple actors' system spanning different levels of governance (Candel, 2014, p. 590). The actors involved in this governance framework encompass supranational, national, and local institutions, private corporations, international organizations, and civil society groups. It is widely recognized that 'food security is a highly complex issue that the current fragmented institutional architecture cannot effectively address' (Candel, 2014, p. 596). Given these challenges, advancing theoretical and methodological approaches to food security governance is essential. A comparative perspective on the interactions and collaborations between public and private actors—along with the engagement of food aid recipients—can provide valuable insights for improving policy outcomes and strengthening governance frameworks.

Given its inherent wickedness, let us consider a working definition of food security governance that combines the contents and procedural aspects. It can be understood as: '*The formal and informal interactions across scales between public and/or private entities ultimately aiming at the realization of food availability, food access, food quality, food utilization, and their stability over time*' (Candel, 2014, p. 598). Building on Candel's definition, one must differentiate between the substantive and procedural aspects of food security governance. The substantive aspect is about ensuring that food is available, accessible, and of good quality, while the procedural side is about the coordination, decision-making, and institutional efforts that drive these outcomes. The substantive aspects incorporate all four FAO content dimensions—*utilization, accessibility, availability, and stability*—to which this work adds the *nutritional status*, linked to a variety of rural and urban programs specifically aimed at improving child and adolescent food consumption at school and within households (Gross et al., 2000). The multifaceted contents of food insecurity rely on the human right to food and healthy food. It involves what is known as the food system, which goes *from farm to fork*, from agricultural/industrial production to the commercial dynamics that affect relationships between producers and consumers. Moreover, it considers access to vulnerable segments of the population (elderly and young people, immigrants, and specific populations that are targeted in specific crisis conditions).

The procedural aspects of food security governance encompass institutional, organizational, and inter-organizational/relational dimensions (Table 1.1), which rely on various key resources: rules, time, and coordination involving various types of actors (public and private) in both formal and informal settings. The procedural aspects of governance encompass how actors publicly commit to collaboration and engage in activities across the various phases of the food security cycle: planning, management,

Table 1.1 Food security governance: substantive and procedural dimensions

Substantive food security governance dimensions	*Procedural food security governance dimensions*
Nutrition status: State of nourishment and the *balance* between food intake and the body's needs for growth	Multiscale interactions (local, national, and international levels)
Availability of food: Physical presence of food in sufficient quantities	Public and private interactions in role definition and responsibility allocation along the food supply chain
Access to food: Adequate resources (entitlements) for acquiring appropriate food for a nutritious diet	Formal and informal operational framework: Establishment of a legal and operational framework
Utilization of food: The body's ability to absorb and metabolize nutrients. It includes food safety, hygiene, clean water, and healthcare	Procedural flexibility: Governance procedures dynamically adapt to environmental change and challenges
Stability of food: The consistent availability, access, and utilization of food over time without sudden disruptions	Monitoring and evaluation (track the progress and measure outcomes, ensuring stability over time)

Source: Made by the author based on FAO (2006), Gross et al. (2000), Candel (2014)

delivery, monitoring, and follow-up. They involve a time dimension, focusing on the stability of these interactions over time, which signals the institutionalization of collaboration within food security programs. Effective procedural governance requires internal coordination methods where traditionally siloed bureaucracies must engage in collaborative knowledge-building, requiring greater interoperability (Sancino, 2022; Sancino et al., 2018). Additionally, it requires external collaboration practices to effectively achieve outcomes that meet citizens' food security needs.

Food security is marked by complexity across both substantive and procedural dimensions. Substantively, it is shaped by interdependent economic, environmental, and social factors that make ensuring consistent access to nutritious food a persistent challenge. Procedurally, governance is fragmented across multiple levels, requiring coordination among entities with differing priorities, regulations, and capacities. The wickedness of food security governance is further manifested in the persistent uncertainty and shifting contexts that shape it over time. Measuring success is difficult: multiple, sometimes conflicting, indicators are used to judge when the problem is "solved," and policy effects range from safeguarding the universal right to food to protecting market interests.

Food security governance combines substantive dimensions (the 'what') and procedural aspects (the 'how'). What is less self-evident, however, is the critical role of the 'who'—those who intervene in the process of making the food security program effective in normal times, such as in times of crisis. One must clearly identify who is responsible for allocating resources, setting objectives, and procedures for implementing, managing, and adapting programs to specific contexts. In some regions and countries, for example, in Latin America, the success of food security programs largely depends on multilevel public intervention for the adoption and implementation of a mix of federal programs based on Conditional income transfer mechanisms to guarantee at the local level access to food, health, and school services (Ferrario, 2014). In European countries, food security intervention depends on collaboration across multiple levels of agencies and the synergy between public and private actors involved in policy formulation, resource allocation, local implementation, and oversight.

The 'who' argument in food security governance is crucial to recognizing the intricate roles that public actors and agencies play in shaping the outcomes of food security initiatives and understanding the complex, multi-layered relationships that emerge at different levels of governance in comparative perspective (Candel & Daugbjerg, 2020). The next matter to consider is the key theoretical perspective of street-level bureaucracy that helps explain how public servants and local-level actors (such as social workers, health professionals, and local government officials) navigate the challenges of poly-crises. These actors often make critical decisions in uncertain and high-pressure environments, directly contributing to shaping governance. Food security governance has descriptive, prescriptive, and predictive implications. From a descriptive standpoint, it provides an empirical analysis of the current procedural tools activated by public and private entities across various geographical scales. This helps to identify how food security is managed by street-level bureaucrats and operators in practice, highlighting the dynamics, challenges, and outcomes of existing programs. The prescriptive aspect of food security governance functioned as a normative guide, offering recommendations on how food security governance ought to ideally operate (McKeon, 2015). This includes advocating for more inclusive, sustainable, and coordinated efforts, ensuring that the right to food is upheld, and that food availability, access, quality, and stability are achieved and maintained over time. In this context, collaboration across public, private, and civil society sectors is essential, as it promotes the sharing of knowledge, resources, and expertise. Street-level bureaucrats play a crucial role in fostering collaboration and acting as intermediaries between policy strategies and local contexts in response to specific security circumstances (Maynard-Moody & Musheno, 2000; Meyers et al., 1998). Finally, the predictive aspect of food security governance focuses on anticipating future challenges, trends, and outcomes. This involves analyzing knowledge from past, present, and potential future crises as a foundation for planning and decision-making. In an ever-changing environment where new food security challenges constantly arise, collaboration and collective learning are essential for developing adaptive policy solutions.

The book integrates existing claims from multiple research fields on the link between collaborative governance, street-level bureaucracy, and collective learning into a consistent framework of causal mechanisms (Capano et al., 2019; Newig et al., 2018) to capture the governance dynamics and the evolving adaptive responses to communities' food security needs during crises (Candel et al., 2016; Lilleker & Stoeckle, 2021; Moon, 2020).

1.2 Collaborative Governance and Food Security Crises: Opportunities and Challenges

Governance involves the challenge of steering collaboration through 'relatively decentralized, network-based interactions among constellations of public and private actors' (Deutsch, 1963, p. xxvii; Gilardi & Radaelli, 2012, p. 155). The steering process develops through substantive and procedural elements that promote problem-solving, reflexivity, and potentially collective learning (Gilardi & Radaelli, 2012, p. 155). Collaborative governance is understood as the processes and structures of public policy decision-making and management that involve stakeholders across public agencies, government levels, and the public, private, and civic spheres, to achieve public goals that cannot be accomplished alone (Ansell & Gash, 2008; Emerson et al., 2012; Greenwood et al., 2021; Lizzi & Righettini, 2022). It is empirically defined by how it is practiced in real-world settings (Greenwood et al., 2021), specifically in this work's case, the provision of food aid programs (FAPs) during crises.

Collaborative governance provides a powerful lens for investigating food security as a social 'creeping crisis,' featuring a gradual, often unnoticed progression and lack of a clear beginning or end (Boin et al., 2021). Disruptive crises such as pandemics or wars act as catalysts, intensifying the vulnerabilities of already struggling populations in need of immediate, innovative responses.

Large-scale crises can influence the structure and dynamics of actor collaboration and the ability to respond to emerging challenges. During a crisis, decision-makers are urged to reach agreements and collaborate on understanding a problem and solving it quickly. They are supposed to drive, manage, and improve existing programs in 'real-time,' going beyond the organizational or professional perspective of conflicting 'discretion' in delivering public services to their clients (Davidovitz et al., 2021; Gofen & Lotta, 2021; Gofen et al., 2019) and of ideological conflicts with politicians (Bertelli, 2021; Lotta & Marques, 2020) when implementing policies (Bertelli, 2006; Stensöta, 2011). Investigating *collaborative efforts* in food security may enable a better understanding of the more or less institutionalized strategy and tools (Ansell & Gash, 2012; Dickinson & Sullivan, 2014; Emerson & Gerlak, 2016; Emerson & Nabatchi, 2015; Klijn & Koppenjan, 2016; Scott & Craig, 2017) for incorporating private actors and citizens' associations into policy processes, steering, mediating, supporting, and facilitating better problem-solving, for a more effective outcome (Agranoff, 2006; Ansell & Gash, 2008; Capano et al., 2015). In food security policies, such an institutionalization means creating policy arenas where 'public actors promote and facilitate, in a shared way, public and private actors' collaboration in pursuing common objectives around a shared problem definition' (Righettini & Lizzi, 2022). The structure and dynamics of street-level collaborative governance (CG) during crises are supposed to adopt a cross-boundaries dynamic to overcome the drifts of siloed policymaking, to interact and coordinate actions with each other by removing organizational boundaries and integrating sets of tools (Maggetti & Trein, 2019; Trein et al., 2019, 2021).

The capacity of collaboration in governance is heavily influenced by contextual factors, including existing coalitions (Edge & Meyer, 2019), the institutional setting, historical governance patterns, and the nature of the policy issue at hand (Howlett & Ramesh, 2016). Institutional settings shape policy implementation as they determine resource availability, organizations and procedures required to execute programs effectively (Sager & Gofen, 2022). A key distinction in this regard is whether the institutional setting is centralized or decentralized, as this distinction affects how actors coordinate, communicate, and resolve distributional conflicts (McKeon, 2015). Several theoretical questions arise: how do decentralized and centralized institutional contexts influence collaborative governance during crises? What practical challenges and opportunities emerge as collaborative governance adapts to these differing institutional arrangements?

Centralized governance modes, while promoting efficiency and uniformity, often struggle to adapt to grassroots legitimacy and specific needs of local communities (Sorensen & Gudmundsson, 2010). Conversely, decentralized systems can foster stronger local engagement, but they may face challenges related to policy consistency and equitable resource distribution. Such a distinction can lead to different **configurations of collaborative governance, particularly in urban food security.**

While collaboration is crucial in both governance structures during crises, the mechanisms and effectiveness of engagement can vary significantly (Wiskerke, 2009, pp. 376–377). In centralized settings, national or regional governments may lead food security initiatives by setting up broad policies and regulations. However, without localized adaptation, collaborations, and stable face-to-face interactions, these top-down approaches may struggle to meet the specific needs of diverse urban populations (Zeitlin, 2016). Collaborative governance always requires institutionalized mechanisms that bridge the gap between high-level policymaking and local implementation, thus ensuring that interventions are tailored and effective for final beneficiaries. In decentralized contexts, urban food security governance is more likely to be shaped by local governments, community organizations, and market actors, promoting more place-based, adaptive responses. While this can enhance flexibility and stakeholder ownership, it may also result in fragmentation and disparities in **resource distribution** across urban areas. Effective collaborative governance in decentralized contexts must therefore balance autonomy with coordination, ensuring that local initiatives are aligned with broader food security goals, while remaining responsive to local needs. In both centralized and decentralized governance structures, the CG framework captures the adaptable modes of interaction that define food security governance as a bottom-up, horizontally oriented activity (Lelieveldt, 2023). By emphasizing the importance of both institutional arrangements and collaborative mechanisms, the framework supports a more nuanced understanding of how food security governance operates through its agents across various contexts, promoting more effective and inclusive solutions to urban food security challenges (Ansell & Gash, 2008). Given the multidimensional nature of food security challenges—spanning from production and distribution to consumption, recovery, and donation—successful policies hinge on stable, trust-based interactions that promote cooperation, adaptability, and shared responsibility among all stakeholders. By fostering these conditions, governance structures can more effectively address the complexities of urban food security while aligning diverse efforts toward common goals.

The literature (Ansell & Gash, 2008; Emerson et al., 2012) identifies three key procedural characteristics of CG that ought to be systematically integrated in food security policies (Righettini & Lizzi, 2022; Lizzi & Righettini, 2022). First, CG initiatives require a formal and stable agreement that endures over time, ensuring continuity and institutional commitment. Second, they must be oriented toward consensus-building, fostering shared understanding and agreement among stakeholders regarding both policy choices and expected outcomes. Finally, CG initiatives require formal or informal joint responsibility across a diverse range of actors—including public institutions, private sector entities, non-profit organizations, citizens, and communities—which would ensure collaborative accountability in policy design and implementation. The characteristics of collaborative governance (CG) outlined here—formal agreements, consensus-building, and joint responsibility—align closely with Candel's framework, which emphasizes the need for structured, inclusive, and decentralized consensus-driven policymaking.

Ansell and Torfing (2021b) identify co-creation as another core functional aspect of CG. Co-creation is not limited to merely developing solutions; its purpose is also to 'evaluate and define problems and challenges; design and test new solutions; select, implement, evaluate and consolidate the most promising ones' (Ansell & Torfing, 2021b, p. 6). Co-creation dynamics are particularly relevant in crisis contexts, where rapidly identifying challenges and the co-designed responses are not just about reacting to crises, but also about shaping how stakeholders interact, adapt, and innovate together in the face of uncertainty (Sancino, 2022). Collaboration among different actors in co-creation initiatives aims to enhance public value (Calori & Magarini, 2021), which means strengthening the collective capacity to achieve together the desired effects (Barzelay, 2019). These co-creation capabilities depend on stakeholders' willingness to share resources, ideas, values, and aspirations to achieve common goals (Ansell & Torfing, 2021a, 2021b).

Despite the theoretical assumption that policy arenas where individuals engage in stable face-to-face interactions will likely drive changes in outcomes (Ostrom, 2009, 2010; Ostrom & Ahn, 2009), the practical challenges and opportunities for collaborative governance and SLBs learning in different institutional settings during food security crises remain unexplored. Additionally, little is known about how street-level collaboration depends on accessible procedural tools that enable continuous adaptation to evolving social demands, as well as where these collaborative adaptations occur to effectively address the needs of end beneficiaries.

1.3 Street-Level Bureaucrats: From Routine Operations to Crisis Response

In recent decades, the environment in which SLBs operate has undergone profound changes owing to multiple disruptive crises. This required them to shift from routine administrative operations to adaptive, crisis-responsive governance (Lacchei, 2024). Street-level bureaucrats (SLBs) mean frontline officials—such as social workers, health operators, teachers, and police officers—who are directly responsible for implementing policies, being on the frontline of public service delivery, and administering programs that enable intermediate and end beneficiaries to access essential tools and resources (Lipsky, 1980). Lipsky uses the term SLB to describe public service workers who interact directly with citizens in their work and who have substantial discretion in executing their work (Lipsky, 1980). Street-level bureaucrats (Arnold, 2021) are more widespread in policy sectors where the state delivers services to the public, such as (Aviv et al., 2021; Brodkin, 2012, 2021; Durose, 2011; Lipsky, 1980): 'schools, policy and welfare departments, lower courts, legal services and other agencies whose workers interact with and have wide discretion over the dispensation of benefits or the allocation of public sanctions' (Lipsky & Hill, 1993, p. 389). SLBs are at the core of the implementation processes, and owing to their close relationships and negotiations with beneficiaries and citizens, they 'socialize citizens to expectations of government services' (Lipsky, 1980, p. 4) reinforcing their role in building trust-based relationships between public administration and the people they serve. Ultimately, 'policy implementation comes down to the people who implement it,' and SLBs ought to be recognized as part of the 'policymaking community.' The decisions SLBs make, the routines they establish, and the strategies and tools they develop to navigate uncertainties and work pressures ultimately shape the public policies they carry out and the quality of the public service they deliver (Brockmann, 2017; Lipsky, 1980). As Hupe and Hill suggest, street-level bureaucrats (SLBs) often acquire unique and irreplaceable knowledge during implementation by drawing from individual and organizational performance levels. The literature highlights the factors that influence their discretion in delivering public services and help them in daily interactions to evaluate a client's level of deservingness (Tummers & Bekkers, 2014; Davidovitz & Cohen, 2022a, 2022b). As key actors in policy implementation, SLBs navigate the intersection between top-down directives and on-the-ground contexts. In doing so, they make discretionary decisions that shape policy effectiveness (Maynard-Moody & Musheno, 2000, 2012, 2022). Street-level bureaucrats operate within a complex web of relationships that vary across different contexts. The concept of governance introduces an additional dimension, as it highlights the multi-layered nature of their action (Hupe & Hill, 2007, 2019). It offers a separate perspective on how street-level bureaucrats collaborate and are held accountable within interconnected organizations. Their ability to collaborate, learn, and adapt to crises is crucial to ensure that policy initiatives remain responsive to changing social demands (Davidovitz et al., 2021). Oversight and flexibility are crucial in ensuring a program's efficiency, especially when, during crises, the swift adoption and implementation of new procedures and practices may determine the success or failure of security interventions (Lotta et al., 2023; Davidovitz et al., 2021).

Studying street-level bureaucrats' performance, also in normal times, remains challenging because their actions are often difficult to define, observe, regulate, and measure. This challenge is further intensified by their frequent operation in environments that external supervisors cannot easily access. As a result, service recipients are often the only ones who can directly witness, evaluate, and provide feedback on how SLBs exercise discretion in decision-making processes and service delivery. During crises, however, SLBs become more visible, especially when analyzing urban health and food crises and the various dynamic forms of social accountability they maintain toward clients and citizens (Chiwona-Karltun et al., 2021). Given the extensive literature on SLBs in routine governance, it is striking how little research has been conducted on their role during crises. This gap is particularly notable because, in disaster and crisis response (Henderson, 2014), SLBs must continuously learn, adapt, and make both major and minor decisions, applying existing rules while interpreting them to fit specific circumstances (Tummers & Bekkers, 2014).[1] External pressures shape how rules are applied at the street level (Hupe & Hill, 2007). In immigration crises, for example, asylum judges face strong managerial pressure to expedite cases, driven by political demands for efficiency (Lacchei, 2024). External pressures also raise critical questions about how crises reshape SLB decision-making in food security policies, alter organizational structures, and influence frontline discretion differently than in non-crisis settings. Recent studies, particularly those examining the COVID-19 pandemic, have extensively explored these issues. They highlight the fact that crises affect not only bureaucratic discretion and professionalism through the digitization of benefits access, but also the well-being of street-level workers and their ability to deliver services amid evolving challenges (Gofen & Lotta, 2021; Davidovitz et al., 2021; Brodkin, 2021; Busetti & Righettini, 2023). However, a critical gap remains in understanding how crises reshape SLBs' interactions with society and beneficiaries beyond professional and organizational dimensions. While recent studies have focused on SLBs' value orientation (Vike, 2018; Cohen & Hertz, 2020) and on their procedural and administrative adaptations, less attention has been paid to how crises alter the social dynamics of service delivery, trust between bureaucrats and the public, and the role of discretion in responding to beneficiaries' needs. Exploring the critical yet often overlooked role of street-level bureaucrats in policy implementation is essential to understanding how crises influence not only policy outcomes but also the direct experiences of those relying on public services (Thomann, 2015). Their role as 'crisis responders' and intermediaries between the government and the public is particularly apparent in food aid programs that use voucher systems. These programs empower citizens by granting them direct access to food services or products. SLBs enhance users' and clients' ability to assess the quality of food products and services (Lodge, 2015; Stirton & Lodge, 2001). Understanding governance changes during crises requires expanding research to examine how program implementation and public service delivery—also in the digital environment (Malandrino & Sager, 2021)—can foster new partnerships between the public and private sectors, which includes the third sector in service co-creation and delivery processes. The take-up system of a food security program is always a lens to closely observe how local decision-makers design their activities and tools (Aizer, 2007; Bergen & While, 2005; Brodkin, 2012, 2021; Durose, 2011) to identify and reach the ultimate beneficiaries, satisfy their needs, and, therefore, align the supply of benefits with the actual demand (Kleven & Kopczuk, 2011). A take-up system means a set of screening, targeting, and delivery activities.

Turbulence captures the increasingly volatile context in which street-level managers and operators find themselves (O'Toole Jr et al., 2005; O'Toole Jr & Meier, 2006), and shatter the old Weberian dogma of bureaucracy as an organizational form for stability and a driver of standard operational procedures. In times of crisis, swift and decisive action is required, and crises are increasingly defined in the plural, using the term 'poly-crisis,' which indicates multiple, interconnected crises—economic instability, climate change, pandemics, and social unrest—that occur at the same time, with each one exacerbating the others (Mark et al., 2023; Moloney et al., 2025; Charbonneau & Giguère, 2025). The interconnectedness of modern challenges means that decisions in one domain inevitably ripple into others, underscoring the need for a holistic and coordinated approach to governance. Traditional, linear crisis management strategies are inadequate for addressing the complexity of these interlinked issues. Instead, adaptive governance and resilient systems are required to maintain the capacity to respond effectively to continuous change over time (Sager et al., 2014). The concept of poly-crises is particularly relevant today, where the interdependence of chaotic and unpredictable events results in compound risks that challenge traditional governance models.

Furthermore, the attention to the growing cross-border crises has been accompanied by an equal understanding of the need to build more adaptive governance dynamics and new individual and organizational capacities such as reflexivity, responsiveness, resilience, revitalization, and rescaling (Boin & Lodge, 2016). In this context, SLBs play a crucial role as frontline responders to poly-crises (Brodkin, 2021; Gofen & Lotta, 2021), as they translate policy decisions into tangible actions that directly affect communities. Their ability to interpret, adapt, and implement policies under high-pressure conditions determines the effectiveness of crisis responses. Given the increasing turbulence in urban contexts fueled by racial, social, and economic conditions, rethinking SLB roles in crisis governance is essential (Brodkin, 2021). In times of emergency and crisis the idea of a program implementation that is guided and overseen with a top-down approach is inadequate. Therefore, multi-agency collaboration strategies are making room, in which horizontal lines of collaboration grow (Klijn & Koppenjan, 2016), where the integrated teamwork of the SLBs is activated (Emerson et al., 2012). Successful implementation in crises and emergencies follows a real-time backward mapping logic (Elmore, 1979), in which the SLB proceeds by making various attempts and successive redesigns of the procedural tools. This evolutionary process features real-time interaction between public, private, and target populations. The COVID-19 pandemic is a unique opportunity for comparative investigation into how policy improvisation, risk exposure, and expanded discretionary power (Davidovitz et al., 2021) shape the role of SLBs across different governance contexts. As frontline implementers, SLBs face heightened expectations from decision-makers, while simultaneously grappling with the limitations of pre-existing service delivery routines (Davidovitz et al., 2021; Busetti & Righettini, 2023). Furthermore, they are crucial in organizing network responses at multiple levels (Henderson, 2014). However, the author argues that the extent to which they were able to adapt, innovate, and learn varied significantly depending on **institutional settings, governance structures, and resource availability**.

During periods of turbulence and crisis, being able to swiftly identify policy targets, reach beneficiaries, and achieve food security outcomes depends on the SLBs' ability to learn and adapt procedural tools in real time (Lodge, 2015; Lodge & Wegrich, 2014). While street-level bureaucrats and frontline operators play a crucial role in managing poly-crises, there is still little research available on the governance dynamics that shape their decision-making and learning in addressing vulnerabilities and food risks in urban settings (Davidovitz et al., 2021; Gofen & Lotta, 2021).

1.4 Bridging the Gap: Policy Learning and Street-Level Practices

Pressman and Wildavsky noted: 'The individual who fails to learn from experience is forever lost in a chaotic world. But learning from experience is not easy… it requires attention to detail' (Pressman & Wildavsky, 1973, pp. 125,126). They underscore that learning is not merely passive but rather an active, deliberate process. It involves adjusting the goals or techniques of a policy in response to new experiences or information. The SLBs' capacity to learn during turbulent times is crucial, as it enables the continuous honing of their approach. This capacity translates into the ability to redesign, update, and adapt intervention tools and the policy network, namely the system of relations between public, private, and civil society entities. The aim of this is to address the new complexity and urgency of emerging problems (Ansell et al., 2021b; Mayne et al., 2020; Scognamiglio et al., 2022). Being able to learn in a complex system or organization is linked to the ability to handle 'the burden'—the influx of messages and signals impacting individuals' decision-making abilities or oversight organizations (Deutsch, 1963, p. 41). This capacity is crucial when the need to manage wicked problems and multiple crises becomes 'the new normal.' Differential learning forms are associated with different fields (policy-specific), and what is learned can include new shared ideas and/or new strategies or actions among the actors within an organization (Knight, 2002). Public management scholars define learning both as a process and an outcome. Those who define it as a process view it as behavior patterns, like reporting errors or asking for help (Tamuz, 2001). Those who define it as an outcome view it as a change—under pressure—in knowledge, actions, and routines (Ellis & Shpielberg, 2003), or both (Dekker & Hansén, 2004).

Policy learning encompasses various mechanisms within the realm of public administration, including lesson drawing, policy diffusion, and policy transfer. It is inherently complex and multifaceted, operating across multiple actors, dimensions, and systems (Bennett & Howlett, 1992). In ordinary circumstances, policy learning is deemed socially desirable, as it aims to improve policy outcomes by adopting effective strategies (Moyson et al., 2017). However, during crises, the need for policy learning becomes even more critical to overcome new challenges and unexpected problems for which there are no ready-made solutions or prior experiences; SLBs may sometimes need to be more aware of the impact of learning when questioned, and they may also engage in policy learning processes without being fully aware of it. 'Learning by doing' at the street level encompasses designing program features that consider 'the relations between givers and recipients under stress conditions' (Pressman & Wildavsky, 1973, p. 136), adopting practical and consistent tools, and ensuring the monitored allocation of resources based on anticipated outcomes. It also requires a commitment to equity and recipient satisfaction. In today's context, this approach also includes creating feedback loops with users through public networks, platforms, and apps. Crises can act as powerful 'learning triggers' that create windows of opportunity to address new situations and expose flaws in existing procedures (Deverell, 2009). However, institutional and political factors within the policy domain (Birkland, 2006; Boin & Rhinard, 2008), as well as cognitive biases of street-level bureaucrats (SLBs), can hinder this learning process and contribute to future inertia (Meyers & Nielsen, 2012). Additionally, the short timeframe and the nature of contingent learning during a crisis may limit the depth and breadth of the knowledge gained once the crisis has come to an end (Busetti & Righettini, 2023).

Crises pose unique challenges for policymakers, requiring rapid and flexible responses. This highlights the urgent need for policies that prioritize accessible and adaptable learning environments (Boin et al., 2018; Meyer & Kunreuther, 2017). As already remarked by Pressman and Wildavsky, 'errors (or successful implementations) can be explained by structural features of relations between aid givers and recipients' (136).

The influential definition of Levitt and March (1988, p. 320) suggests that learning occurs when organizations establish new routines. They define routines as 'beliefs, frameworks, paradigms' as well as 'forms, rules, procedures' (1988, p. 320). This definition is shared with other policy scholars as it encompasses knowledge and behavior and captures knowledge acquisition and behavioral adaptation (Gerlak & Heikkila, 2019). It acknowledges that learning goes beyond mere knowledge acquisition; it involves integrating new information into established routines and shaping organizational behavior patterns and actions. By emphasizing changes in routines, the managerial perspective on learning centers on the internal dynamics of governmental processes within organizations, focusing on the interactions between bureaucrats and operators. The perspective completely explains how learning takes place within the organizational context. However, another source of collective learning comes from the interaction of organizational members with the external environment, recipients, stakeholders, and beneficiaries of programs and services.

The recognition of a double-loop learning in organizational contexts (Auqui-Caceres & Furlan, 2023) highlights the distinction between two learning pathways: the internal in-in pathway, which focuses on internalizing knowledge, experiences, and data to hone decision-making, and the in-out pathway, which integrates external insights to enhance adaptability and responsiveness. The 'in-out' pathway differs from the 'in-in' pathway by focusing on the outward application and sharing of knowledge. While the 'in-in' pathway centers on internal knowledge processing and refinement, the 'in-out' pathway actively engages with the external environment—applying, scaling, and disseminating learned knowledge. Its relational aspects are crucial, requiring organizations to integrate and process insights from various external sources, including what are known as the 'non-expert'—clients, users, intermediate and end beneficiaries, third parties, and stakeholders. Effective organizational learning depends on the ability to assimilate and employ information from both internal (professional-expert) channels and external (common non-expert) channels, leveraging best practices, research findings, and feedback to drive continuous improvement. SLBs must actively assimilate and apply insights from both internal and external sources to implement public policies effectively. Organizational learning is enhanced by external inputs, particularly from intermediate and end beneficiaries. Traditional Weberian *iterative* learning, grounded in expertise, routines, rules, and the law is increasingly challenged by *interactive* learning, which relies on collaborative exchanges with non-expert human feedback to navigate complex tasks more effectively. While the first type of learning tends to be backward-looking, relying on established knowledge, practices, and routines, interactive learning is forward-looking, dynamic, and responsive, as agents quickly adapt to changing conditions and unforeseen challenges by adopting strategies aimed at meeting goals (Simon, 1947), and engaging in 'mutual adaptation' to evolving environments (Browne & Wildavsky, 1999, p. 219). During crises, the effectiveness of policies depends on dialogue and feedback from both experts and non-experts, including beneficiaries (Cairney & Wellstead, 2021) to ensure that programs and services remain responsive to rapidly changing needs and public perceptions (Janssen et al., 2021; Millard et al., 2022).

1.5 Learning from the Unexpected: Collaborating with 'Non-Experts'

The citizen-agent perspective represents a key shift in how we understand the role of street-level bureaucrats (SLBs). Rather than acting as mere rule-following state agents, SLBs are viewed as moral interpreters and active agents, responding to citizens not only through policy dictates but also by interpreting individual situations through their own ethical frameworks and professional judgment (Maynard-Moody & Musheno, 2022). This approach has been central to illuminating the discretionary nature of frontline work and the relational processes through which SLBs construct and enact their roles. However, the learning framework remains largely unexplored in terms of public service food delivery such as school canteens, and food security programs, particularly in their street-level implementation (Busetti & Righettini, 2023). However, the capacity to learn from non-experts is valuable and transformative. Harnessing the direct experiences and practical insights of those on the ground can redefine how policies are shaped and implemented, ensuring they are more responsive and effective

> *The man who wears a shoe knows best that it pinches and where it pinches, even if the expert shoemaker is the best judge of how the trouble can be remedied.* (Dewey & Rogers, 2012, p. 224)

Likewise, people experiencing food insecurity have the clearest picture of their struggles—whether it is difficulty accessing nutritious food or affordability issues. For instance, a community surrounded by food deserts is likely to have the best grasp of where they lack access to healthy food, even though experts can help them find ways to address these issues, for example, by introducing food markets or improving transportation (Smith & Thompson, 2023). There can be a tension between expert-driven and practice-based learning. While expert-driven learning is more structured and confined to organizational boundaries and can lead to efficient, theoretically sound policies, practice-based learning offers the adaptability and responsiveness needed to address the complexities of the real world, particularly in crisis contexts.

To fully grasp this dynamic, one must understand the shift from collective learning (Heikkila & Gerlak, 2013) to *collaborative learning* (Dillenbourg, 1999; Walker & Daniels, 2019). A concept based on the idea that public and private actors in controversial public policy situations look for 'what works' in the field (Walker & Daniels, 2019). While collective learning remains confined to inter-organizational knowledge exchanges, collaborative learning goes beyond these boundaries by including conflicts with beneficiaries. In collective learning, decision-making is typically concentrated within a group of experts, who exchange knowledge to interpret and apply new information effectively. In contrast, collaborative learning emphasizes the inclusion of both experts and non-experts, encouraging co-creation, innovation, and adaptability in policy development. Collaborative learning emphasizes mutual responsibility, co-creation of interventions, and trust building among various types of actors—including governmental agencies, non-governmental organizations (NGOs), private sector entities, grassroots organizations, community leaders, users, and consumers. However, the nature of mutual trust in collaborative settings varies significantly depending on the stakes, reciprocity, and institutional frameworks associated with different SLB professional roles. For instance, in law enforcement, trust is directly tied to life-threatening situations. Police officers' willingness to risk their lives depends on the extent to which they trust their peers, managers, and institutions (Cohen et al., 2024). Distrust may result in stricter bureaucratic checks, potentially delaying or limiting assistance during critical times. While SLBs in food aid programs do not face any physical danger, their trust in recipients influences aid distribution. Here, trust is not about taking personal risks, but about reducing the risk that others may be excluded from accessing benefits. This approach fosters collaboration and helps minimize inefficiencies. Unlike in law enforcement, where trust issues may arise through social biases, trust in food aid policy is primarily linked to effective and timely service delivery (Cohen et al., 2024). Similarly, SLBs' emotional resilience and coping strategies, in crisis situations, can have severe consequences on public welfare as they may exacerbate vulnerabilities and hinder effective collaborative emergency responses (Maynard-Moody & Musheno, 2000; Davidovitz & Cohen, 2022a, 2022b).

Collaborative learning builds bridges and connects policymakers with those on the ground, therefore, a decentralized and open learning environment tends to be far more receptive to diverse perspectives and better equipped to acquire and interpret information compared to a more rigid, centralized structure. By encouraging broad participation, such structures create space for non-experts with direct experiences and on-the-ground insights to make a major contribution to the learning process. Social dynamics within *collaborative learning* are equally vital. Trusting relationships, open and transparent communication support, and developing shared languages create an environment where dialogue can thrive. These elements enable diverse actors to engage in meaningful exchanges, breaking down barriers between experts and non-experts. This inclusivity not only enhances the sharing of information but also strengthens collaboration. Groups begin to recognize and appreciate the value of insights rooted in real-world experiences and unconventional viewpoints—perspectives that may otherwise be overlooked. By breaking down traditional hierarchies of expertise, collaborative learning becomes a powerful tool for generating creative, adaptable solutions to complex challenges, grounded in the direct experiences of all participants.

Although the original Heikkila and Gerlak's framework (2013) is well-founded, it overlooks crises as a critical external factor that can significantly impact learning. Time constraints often lead to what Kamkhaji and Radaelli (2017) describe as contingent learning, where street-level bureaucrats (SLBs) initially rely on existing networks, rules, and pre-established practices to respond, only engaging in meaningful learning and adaptation after the immediate crisis has passed (Barrett, 2004). Decision-making becomes reactive rather than reflective, often prioritizing speed over deliberation.

During a crisis, SLBs discover new criteria to allocate decisions (Pleysier & Kilkelly, 2024), new sources of learning in the implementation process as they are compelled to continuously adapt and redefine their relationships with both beneficiaries and partners. This dynamic process requires them to reshape learning processes and outputs—such as strategies, plans, programs, rules, services, and procedures—to address the evolving social needs of the end beneficiaries (Howlett, 2020). In a context where material conditions shift rapidly, distributive justice is understood as a global principle (Diaz-Granados et al., 2023) deal with tensions between operational demands and social equity goals (Maynard-Moody & Musheno, 2012). During crises, equity needs to be reframed as the substantive (democratic) and procedural (pluralistic) guarantee (Tyler, 1988) that the most vulnerable populations and groups can access and benefit from essential aid (Tschersich & Kok, 2022; Puupponen et al., 2023). Collaborative learning hinges not only on formal entitlements but also on the real-world mechanisms that determine whether aid reaches those who need it most. Building on the seminal articles by Gerlak and Heikkila (2011, Heikkila and Gerlak (2013), this work's argument is that collaborative learning in food security governance harnesses the power of experts and non-experts through the beneficiaries' formal and informal feedback to guide policymakers through complex challenges, especially during turbulent times. By integrating diverse insights, collaborative learning enables decision-makers to continuously hone their understanding, improving the effectiveness and relevance of their actions in real-time. Our approach prioritizes a 'street-level' perspective in the processes involved, where community members' firsthand experiences and insights are vital in shaping decisions and practices. The bottom-up feedback loop enhances adaptive decision-making and ensures that policies are responsive to the evolving needs and expectations of the communities they serve. This model positions collaborative learning as a dynamic, community-centered process for resilience building in food security governance.

Collaborative Learning (CL) in food security governance is a process where multilevel public and private actors leverage procedural tools and formalized and informal interactive practices to acquire, interpret, and disseminate knowledge from and toward diverse sources, including *experts*—colleagues, professionals, and *non-experts*—intermediate and final beneficiaries—to achieve more satisfying and informed outcomes.

When effectively integrated, Gerlak and Heikkila's (2011) analytical framework offers a powerful lens for understanding where and how knowledge is collaboratively generated, and by whom it is shared and applied in street-level food security governance during a crisis. The framework will enable the division of the knowledge process within the Food Aid Programs (FAPs) into three distinct yet interconnected collaborative steps: information acquisition, translation, and dissemination.

- The *acquisition of information* during crises involves gathering comprehensive insights and data on new and unexpected community needs from various sources (internal and external) regarding emerging food security issues and putting them into the policy agenda. In this phase, all public and private actors are engaged formally or informally in co-designing the food aid program in response to a complex crisis.

- T*ranslating information* enables decision-makers to understand and integrate new information into their decision-making processes. However, during a poly-crisis, this phase becomes particularly challenging, as crises create recurring high-pressure environments, increase uncertainty, and require rapid responses to urgent, challenging problems. In these circumstances, translating information into actionable and socially sustainable procedures requires quick thinking, inter-organizational and inter-crisis learning, and coordination. Inter-organizational and inter-crisis learning (Busetti & Righettini, 2023) requires a formalized yet adaptable relationship between street-level bureaucrats, intermediate and ultimate beneficiaries, and current and potential aid recipients. Mechanisms such as program in-in and in-out feedback loops, which can either be positive or negative, and foresight capacity play a crucial role in this context. By providing rewards or recognition from non-experts for desired behavior patterns or outcomes, these mechanisms reinforce and encourage the continued application of experiential knowledge. This approach helps policymakers and practitioners support effective practices by enhancing behavior patterns that contribute to resilience and informed decision-making in crisis conditions.
- The *dissemination of information* acquired in the context of poly-crises, where multiple, interconnected crises occur simultaneously (e.g., economic, environmental, political, and health crises), faces significant challenges, particularly when it comes to stabilizing the flow of information over time. A poly-crisis increases complexity and uncertainty, complicating how information is shared and processed within organizations and various levels of government or institutions. As something that is learned becomes institutionalized, the latest knowledge or skills are incorporated into individuals' routines and become part of their repertoire. This integration ensures that learning is sustained and integrated into organizational routines.

Collaborative learning considers formalized and informal interaction dynamics and cooperation among SLBs and diverse stakeholders during crises to co-create outputs and redesign timely processes. Collaborative processes often involve a wide range of participants—including public and private actors, experts, volunteers, NGOs, non-profit organizations, and end beneficiaries—working together to address wicked problems. While collaborative learning is relatively straightforward to trace within a single policy field and during stable conditions, its application across diverse policy sectors—especially in times of crisis—highlights its increasing importance in fostering responsive and innovative policies that can adapt to urgent and complex challenges (Lang & Brüesch, 2018, 2020; Lang & Stoeger, 2018; Linke et al., 2022).

Street-level bureaucrats (SLBs) foster open networks when they engage in a continuous cycle of learning, applying, receiving feedback, and honing actions (Moyson et al., 2017). SLBs are expected to continually adapt policy strategies, tools, and techniques to address the changing situations encountered in direct interactions with affected communities (Bennett & Howlett, 1992; Gerlak & Heikkila, 2011; Wiskerke & Viljoen, 2012). Through these real-time adjustments, SLBs contribute to a responsive and adaptable food security governance framework, thereby improving crisis management and enhancing resilience.

1.6 Connecting the Dots: Mechanisms Driving Collaborative Learning

A fundamental understanding of the mechanisms underpinning street-level collaborative learning is essential to grasp the evolution of food security governance during crises in the interactive dialogue and trust within and outside administrative organizations (Lotta et al., 2023; Maynard-Moody & Musheno, 2012, 2022). Turbulent times and unfamiliar circumstances may require swift and adaptive responses (Gofen et al., 2019). In such situations, the collaborative aspect of group work and networks becomes paramount (Heikkila & Gerlak, 2013). The crisis is a crucial pressure test for pre-existing management systems, prompting SLBs to explore and utilize various internal and external strategies. SLBs engage in a learning journey that often begins in response to disruptions and unfolds through phases of information gathering, sense-making, and collaboration.

The core idea behind a mechanistic approach is to open the often-opaque process of adopting and implementing food assistance programs (FAPs) by examining actions and interactions of public and private policy operators on the ground, using the concept of *concatenation of mechanisms* (CM) (Ongaro, 2019) which may be manipulated—triggered of defused—to achieve one's desired goals. Most concrete governance changes 'require resorting to several social mechanisms' that that can either conflict or align (Hedström & Swedberg, 1998, p. 2; Hedström & Ylikoski, 2010). The concatenation concept illustrates 'regular sequences' of social mechanisms that, in the case described in this work, illustrate how crises trigger a chain of decisions and interactions that ultimately shape learning in governance change and outcomes. By tracing a sequence, the CM provides a powerful lens to understand governance development, that is, how the status of food security governance may develop into another state of governance in a turbulent environment owing to crisis pressures. The CM is employed to offer a nuanced diagnosis of where food security policies may thrive or falter, offering insights that are critical for enhancing food security responses in times of crisis.

Table 1.2 outlines the sequence of specific mechanisms that provide a theoretical framework for understanding how street-level policymakers adapt food aid programs to respond effectively to conditions of external crisis. The CM focuses on policy learning phases, that is, structural and procedural adjustments to respond effectively to food security challenges. Table 1.2 highlights how the collaborative learning process in food aid governance unfolds along three interconnected phases applied to actors and actions at supranational, national, and local levels along the governance process. The entire sequence of mechanisms, which will be explained in detail later in this section, is triggered by external crises and provides a descriptive and prescriptive framework for institutional action. The mechanisms collectively contribute to developing the capacity to interpret, manage, and effectively address the food security needs of the most vulnerable populations.

External crises—such as economic shocks, pandemics, or natural disasters—are pivotal moments that reveal the strengths and weaknesses of existing governance structures. These crises are catalysts of street-level exchange of knowledge, ideas, and strategies, which are vital for improving policy responses. Therefore, collaborative learning is a dynamic, iterative process that goes beyond mere information sharing among public and private actors, as it fosters continuous adaptation (Gerlack, Hikkila & Lubell, 2013). The iterative nature of collaborative learning ensures that governance systems related to food security remain flexible and responsive to evolving needs and circumstances. By focusing on systematic reflection—prioritizing the analysis of current interventions over past

Table 1.2 Concatenation of mechanisms (CM) for building collaborative learning and trust from the street level in the food aid program

Crises/turbulences	*Learning phases*	*Equivalent mechanisms of action and trust building (MoA)*	*Behavioral change technique in food security governance*
Increasing food insecurity	Acquisition of knowledge:	(A) Focusing attention	Increasing the attention to and salience of food security by modifying the governance agenda and introducing or integrating new food security purposes and tools
		(B) Frequent interaction and stakeholder certification	Formalizing procedural tools for network coordination and brokerage among different institutional and non-institutional actors—Intermediate and final beneficiaries
	Translation of knowledge:	(C) Target recognition	Co-creating with beneficiaries a take-up system (eligibility criteria) for meeting new needs and demands
		(D) Feedback and backward mapping to extract data on new beneficiaries	Adoption of procedural tools for the revision and adaptation process, based on beneficiaries' reactions, monitoring, and evaluation
Collaborative learning in food aid program (FAP)	Diffusion of knowledge:	(E) Program reuse	Emphasizing the practical advantages of replicating existing policies, driven by the pursuit of efficiency and the desire to capitalize on previous (net)work
		(F) Return to routine and institutionalization	Embedding innovative solutions into standard practice, ensuring they are replicable beyond the initial crisis

practices—governance systems can assess what works within new contexts and determines where there is room for improvement.

A. *The role of crises in focusing attention on food security*. Focusing events are catalysts for agenda shifts and the continuation of agenda activities. Research indicates that focusing events can highlight recurring problems that persist over an extended period and significantly alter power dynamics within a specific policy area (Eckersley & Lakoma, 2022). In recent European history, various food crises linked to financial and environmental challenges—such as droughts and floods (Ortiz-Miranda et al., 2016)—and global events, including EXPO 2015 (Busetti & Dente, 2018), have heightened the focus of public actors on the persistent issue of food security (Campomori & Bassoli, 2024; Melloni, 2024; Righettini & Bazzan, 2017). Over the past decades, recurring food crises, such as the outbreak of mad cow disease, have had wide-ranging effects across multiple countries, shifting governance priority. The frequent and harmful food crises led to the certification of food production processes and enforced coordination of knowledge production and diffusion within the European network (Bazzan, 2021; Silano & Silano, 2020). The pandemic reshaped the value system and transformed public and private actors engaged in policymaking, from problem definition and agenda-setting to data collection, procedural innovation, coordination, and improved information sharing to protect consumers. During the pandemic, governments at different levels unexpectedly prioritized food security issues. The focus on health-related risks and the primary task of managing the pandemic (Taylor et al., 2023) have not overshadowed the significance of secondary effects such as food safety and deprivation. In many countries, the announcement of the pandemic was a window of opportunity for legislative measures that contributed to raising the attention of all local government officials on the issue of access to food and food safety during the health crisis. The pandemic made it once again a central topic, which resulted in a quantum leap in urban policies and, in many cases, an honest and prior institutionalization of politics (Busetti & Dente, 2018). The mechanism of focusing attention on food security can manifest itself in various *scenarios*, each differing in terms of timing, stability, and focus within the institutional agenda.

Sustained Attention on Existing Food Security Policies. In this scenario, governance actors reinforce and amplify pre-existing food security initiatives by mobilizing additional resources and institutionalizing new procedural tools. The focus is on increasing and scaling up ongoing efforts to ensure that food security measures remain robust and effective throughout the crisis and catalyze more profound institutional commitment.

Integration of New Food Security Priorities into Existing Programs. Another possible scenario involves governance shifting its attention to new food security issues that have emerged in light of the crisis. In this case, new priorities are integrated into existing programs and resources. This adaptive response seeks to address emerging needs while building on the foundation of established policies and initiatives. The stability of this approach depends on how seamlessly these new priorities are incorporated and whether they continue to be supported once the immediate crisis has passed.

Temporary Focus on New Food Security Programs. In some instances, governance may develop a temporary focus on new food security programs and resources that are not connected with pre-existing initiatives. This scenario often arises when a crisis creates urgent and unforeseen demands that require immediate action, leading to the rapid deployment of short-term interventions. While these efforts may provide critical relief, their temporary nature and lack of integration with broader governance structures may limit their sustainability and long-term impact.

Each scenario will likely involve different compositions of actors and varying levels of interaction with beneficiaries. For example, strengthening existing programs may involve established public institutions working closely with community organizations and employing familiar engagement channels. In contrast, launching new initiatives may involve a broader range of stakeholders, including private sector partners, NGOs, and international organizations, which would require the creation of new networks and building trust.

2 *Frequent interactions and trust mobilization (stakeholder certification).* Trust can be defined as 'the actors' more or less stable, positive perception of the intentions of other actors to collaborate, that is, the perception that other actors will refrain from opportunistic behavior' (Edelenbos & Klijn, 2007; E.-H. Klijn et al., 2010; Davidovitz & Cohen, 2022a, 2022b). In normal circumstances, SLBs play a crucial role in ensuring that knowledge flows effectively among diverse stakeholders, enabling trustworthy decisions and coordinated action (Maynard-Moody & Musheno, 2022). The citizen-agent model emphasizes SLBs' role in adapting policies based on contextual realities

and individual cases. Serving as a 'citizen agent' means many things: adjusting policies to fit citizens' unique needs, making governance appear more responsive and compassionate, exercising discretion fairly and consistently, and reinforcing the legitimacy of public institutions throughout transparent and fair procedures (Tyler, 1988, Tyler, 2006; Maynard-Moody & Musheno, 2000). However, during crises—when the role of SLBs extends beyond administrative duties to life-risking services—trust becomes a defining factor in their willingness to act decisively for the well-being of others (Cohen, 2022). SLBs, such as healthcare workers, emergency responders, and frontline public servants, must navigate uncertainty, conflicting demands, and heightened public expectations. Their ability to make life-and-death decisions hinges on trust—both in the institutions they represent and in the cooperation of the communities they serve (Cohen, 2022). Given the urgency of responding swiftly and effectively to crises—such as economic crises, pandemics, wars, or geopolitical conflicts—it is no surprise that scientific interest has grown in investigating how trust influences policymaking (Devine, 2024; Verhoest et al., 2025) and street-level decisions in conditions of resource shortage or bureaucratic red tape (Davidovitz & Cohen, 2023; Maynard-Moody & Musheno, 2022; Henderson et al., 2020; Edelenbos & Klijn, 2007).

Trust, as a relational concept, plays a critical role in navigating uncertainty and assessing risks. It shapes both bureaucratic discretion and the effectiveness of procedural tools for coordination and collaboration (Greenwood et al., 2021; Davidovitz & Cohen, 2023). These tools outline how to manage the interdependencies between various actors and their resources involved in the decision-making process (Cucciniello & Linos, 2022; Simon, 2013), and how to achieve a higher level of satisfaction than could be impossible through unilateral decisions and actions alone (Scharpf, 1997). Trust is a key factor in inter-organizational collaborations, enhancing cooperation, reducing conflicts and costs, and facilitating knowledge exchange through informal relationships. It both influences and reflects the success or failure of partnerships and collaborative alliances (Gulati & Singh, 1998; Vangen & Huxham, 2013). It is universally accepted that trust plays a crucial role in fostering cooperation, coordinating efforts and networks, maintaining social order, and minimizing reliance on forceful governmental interventions (Cairney & Wellstead, 2021). Trust matters in collaborative governance networks (Edelenbos & Klijn,

2007; Klijn et al., 2010) and it is achieved by facilitating intense relationships where opportunistic gains are banned by rules that are fair and are shared: 'trust affects the cooperation between the actors in the networks and the flow of information it facilitates between actors for innovation' (Klijn et al., 2010, p. 318). However, emergency service delivery involves not only collaboration but also organizational change, making it essential to consider internal trust dynamics and their impact on strategic partnerships. In Europe, trust-building strategies have been particularly important in recovering from food safety incidents and scandals—requiring negotiations on beliefs and diverse risk perceptions (Ansell & Vogel, 2006; Ma & Christensen, 2019). EU institutions have identified cultural and procedural tools to restore consumers' trust in food safety regulations (Kjærnes et al., 2006). These include transparency, the certification of independence for regulatory agencies (Righettini & Bazzan, 2017), the development of clear protocols and procedures along the food chain, ensuring credibility, being proactive, prioritizing the public's interest, collaborating with stakeholders, maintaining consistency, educating both stakeholders and the public, building a strong reputation by keeping promises (Henderson et al., 2020). Mutual trust enables actors to play a cooperative game in institutional and inter-organizational work (Aumann & Maschler, 1964; Elston, 2017; Majone, 1989; Weiss, 1980). Nevertheless, people experiencing material deprivation and financial hardship are more likely to distrust their national government (OECD, 2024). During crises, and more generally in an increasingly complex policy environment, satisfaction with the provision of key public services and material assistance influence people's trust levels (OECD, 2024). Therefore, policies that promote social inclusivity while addressing economic and material vulnerability and discrimination play a key role in closing trust gaps (OECD, 2024). As Lypsky argued, the day-to-day interactions with public institutions remain key drivers of trust in civil service and local government (OECD, 2024). Building on this insight, a clear conclusion is that *any strategies to build a sense of SLB legitimacy in the eyes of the public need to focus on substantive and procedural aspects* (Hough et al., 2013). The literature identifies two specific administrative capabilities (Lodge & Wegrich, 2014) that can enhance trust in governance under normal conditions (Thomann, 2015; Thomann et al., 2018): firstly, the ability to activate, and secondly to intensify the interaction and collaboration among diverse actors to achieve common objectives through shared strategies (Abramson & Mittu, 2006; Scerri et al., 2006; Scott & Gong, 2021).

Frequent interactions play a critical role in collective learning (Heikkila & Gerlak, 2013) and are central to the capacity for managing wicked problems in turbulent times (Allen et al., 2011; Ansell et al., 2017, 2021b). While trust within a collaboration process is typically developed over time, crises demand its rapid establishment (Mayer et al., 1995). Uncertainty, shifting priorities, and resource constraints can strain relationships, making transparency, accountability, and consistent communication crucial. In turbulent times, trustworthy information and reliable solutions are crucial for a practical navigation process. Bad or inaccurate information can undermine these efforts, harming those relying on food assistance (Deutsch, 1963; Nonaka et al., 2006).

Trust elicited in urgent situations can be quickly granted but it remains fragile and can easily be eroded if expectations are not met, or conditions change rapidly. Trust is more easily established when local actors are perceived as closer to citizens, more responsive to immediate local needs, and more familiar with community-specific challenges (Ansell & Gash, 2008). However, in systemic crises, coordination failures within decentralized systems can **weaken and undermine** the effectiveness and legitimacy of crisis management efforts (Peters & Pierre, 2000; Wild et al., 2023). This is one reason why, as will be seen in the next chapter, during recent crises the European institutions exerted strong intergovernmental coordination, to increase mutual trust and collaboration between citizens and public officials.

3 *Target recognition*. Along the collaborative learning process, identifying and addressing changing population needs is a crucial phase of policy-making, especially amid crises. Target recognition refers to the ways SLBs extract information from program beneficiaries to shape and reshape the communication strategies and the take-up system to meet the most vulnerable groups (Esobi et al., 2021; Radimer et al., 1992; Radimer, 2002) collecting and interpreting data on the evolving context (Lotta et al., 2023; Power et al., 2020). The social and administrative structure of target populations of food security in turbulent times meant that governments needed to act swiftly and thoroughly in response to crises. In some cases, they could not rely only on their organizational resources and coordination, they also needed the support of civil society, in its various organizational forms, to identify the needed action (Henderson et al., 2020). In food aid programs, recognizing social targets refers to the SLB's ability to rapidly understand the causes

of new food poverty and vulnerability and their consequences for different potentially vulnerable groups. At a macro level, new certification tools—such as the AROPE index (people at risk of poverty or social exclusion) in Europe—fulfill this function and provide European Member States with new criteria to measure and govern food insecurity programs at the national and local levels (Cafiero et al., 2018). Different metrics can uncover diverse networks and strategies for identifying and supporting needy individuals. At national and urban levels, straightforward indicators of food poverty, such as income or food expenditure and their charitable dimensions, entail advantages and limitations. A key strength lies in the presence of formal and informal (charitable) assistance networks that rely on a more flexible interpretation of income. However, a major limitation is that defining 'food poverty' solely based on income often overlooks population segments with pressing needs not directly tied to traditional economic classifications. Qualitative investigation and multi-variable analysis, such as the 'vulnerability index', reveal that from a permanent-income standpoint, single-parent households may face increased child deprivation owing to unexpected and disruptive events that temporarily limit their opportunities to share employment risks among adults in the household or to provide adequate food for the family (Action Aid, 2022, 2023). Specific social categories are generally more exposed to vulnerability than others: immigrants, self-employed workers, children, individuals living alone, and unemployed people. For example, European statistics indicate that in certain countries, such as Bulgaria, Estonia, Italy, Cyprus, Latvia, Lithuania, Malta, Poland, Portugal, Romania, Serbia, and Slovakia, a single-level approach to understanding child deprivation is less effective. In these countries, the overall standard of living tends to be lower, which means that children in these areas are more likely to experience deprivation. Therefore, data collection and interpretation require a multifaceted approach, especially during turbulent times, to fully capture and meet the needs of end beneficiaries and achieve effective outcomes (Aaberge et al., 2021).

Several scenarios can manifest themselves when analyzing the recognition of the most vulnerable in urban food security policies, considering *eligibility criteria* to reach those needing help the most. There may be inclusive targeting when programs are designed to capture a broad spectrum of vulnerable populations, with eligibility rules that are flexible and comprehensive. The scenario ensures that various groups receive the proper support, including those with less apparent vulnerabilities. The outcome is a well-rounded and equitable distribution of resources. A selective targeting scenario is characterized by specific, predefined categories of vulnerability, such as low-income households or the elderly, potentially leaving out other at-risk groups such as the newly unemployed or undocumented migrants. This scenario may lead to gaps in assistance, where some vulnerable populations are underserved or overlooked. An overly restrictive eligibility scenario adopts stringent criteria, resulting in a significant portion of the vulnerable population being excluded from assistance. While the focus may be on controlling costs or preventing fraud, this approach often fails to reach those in genuine need, further exacerbating food insecurity. The result is uneven and fragmented protection, where some areas provide comprehensive aid while others fall short, creating a patchwork system that inadequately addresses food insecurity on a broader scale. In response to evolving crises, eligibility rules and target populations are frequently reassessed and adjusted dynamically. During a crisis, a highly responsive and adaptive policy framework is essential to quickly identify and address emerging vulnerabilities, ensuring that no group is left behind.

4 *Feedback and backward mapping*. Scholars view knowledge in policymaking differently. Some believe knowledge is external and directly applicable to policy, while others see knowledge as internally shaped by policymaking. Some scholars emphasize top-down rational strategic planning, organizational resources, and factors such as values, relationships, and policy experts influencing knowledge use in policymaking (Migone & Howlett, 2024). Synoptic, top-down approaches, looking at the implementation gaps, often overlook the role of errors, participation, and evaluation as sources of knowledge (Capano & Malandrino, 2022), particularly all the knowledge resources from the beneficiaries themselves, which can contribute to collaborative learning. In turbulent times, when demand for services is under pressure and remedies are unknown and unpredictable, available organizational resources may act as a constraint rather than an opportunity. Crises reveal SLBs' constraints and weaknesses and their potential to reorganize knowledge resources using face-to-face interaction with stakeholders and beneficiaries (Busetti & Righettini, 2023). The governance of urban food aid programs provides a valuable lens for understanding how street-level

bureaucracy adapts and learns through direct interactions and collaborative practices within policy implementation (Cohen, 2021; Cohen & Aviram, 2021; Zhang et al., 2021) delivery. Feedback mechanisms from individuals and communities directly impacted by the food security program enable street-level bureaucrats to use backward mapping to adapt to real-world challenges. The relationship between feedback from beneficiaries and backward mapping is synergistic. The backward mechanism integrates past feedback (negative or positive) into future policy development through continuous monitoring and evaluation tools, focusing on how outcomes are experienced and achieved at the ground level. Only the interrelation between these two mechanisms ensures that the program design is closely aligned with the situations of the target population, as it involves thinking through how each program element will operate in practice, considering the local context and the needs of the beneficiaries. Backward mapping is a bottom-up mechanism whose purpose is to interpret new information phases to improve the policy outcome. In a crisis, feedback and backward mapping allow for a quick identification of what idea or instrument is working or failing at the ground level of the implementation process. It provides useful information to decision-makers and starts from service delivery. It focuses on the feedback on outcomes and works backward to identify the steps and decisions that lead to those outcomes. It emphasizes the roles and actions of local government officials and field workers who implement the policies. The *bottom-up* backward mapping mechanism is activated by the implementation experience of the people who are directly affected by the program or services, working up to the higher levels of policy design and decision-making. The mapping process traces the output delivery and outcomes and examines what contextual factors led to complete or partial success or failure. A backward mapping mechanism provides SLBs with a unique way to evaluate a short-term urban aid program and the effectiveness of the ongoing adaptation to beneficiaries of the take-up system (Beutel et al., 2023). The dynamics of backward mapping in food security programs are pivotal in identifying gaps and understanding how grassroots actors shape policy outcomes. This approach emphasizes the importance of connecting policy design with on-the-ground implementation, ensuring that insights from the community level inform and hone higher-level strategies. By focusing on backward mapping, one can dissect how local innovations and experiences with monitoring and evaluation systems are captured, analyzed, and employed to improve food security interventions (Carless, 2019; Van Acker & Bouckaert, 2018). Communities and frontline workers, who are often the first to confront food insecurity challenges, provide critical feedback through monitoring and evaluation to adapt and improve programs in real time, ensuring they remain responsive and effective. By systematically gathering data on outcomes and processes at the street level, policymakers can identify gaps, inefficiencies, or areas for improvement that may not be apparent from a top-down perspective. Two scenarios can be observed when applying backward mapping within food security governance. Adaptive feedback loops become crucial in acute food insecurity or crises. Frontline workers engage directly with affected populations, collecting real-time data that is immediately applicable to decision-making. This dynamic process ensures that immediate needs are met while informing future strategies. For instance, data gathered on food distribution bottlenecks or dietary deficiencies can be used to adjust logistics and supplies in real time while contributing to predictive models and planning subsequent response phases. In more stable contexts, backward mapping supports a structured learning approach, where feedback from long-term food security programs is systematically integrated into policy development. Here, the emphasis is on analyzing longitudinal data to understand trends, assess the effectiveness of interventions, and hone policies over time. Community-based monitoring becomes a cornerstone of this process, ensuring that programs remain aligned with the evolving needs of the population and the changing environmental, economic, or social conditions. In a second scenario, traditional monitoring and evaluation systems may prove inadequate to track the effectiveness of interventions in real time and develop forecasting activities. A city may, for example, implement a digital tracking system that monitors the flow of food aid from warehouses to recipients. Insights gathered from this system can reveal inefficiencies, such as delays or gaps in digitization or database integration. In both scenarios, integrating feedback into the policy cycle is paramount. The continuous loop of information flow—from the field to policymakers and back—fosters resilience and sustainability in food security governance, aligning national strategies with local situations for maximum impact. The ongoing learning process creates a cycle of adaptation that shapes the trajectory of food security policies. For example, suppose a food distribution program encounters bottlenecks due to inadequate supply chains. In that case, the feedback from street-level workers can lead to a rapid policy shift, reallocating resources or altering distribution strategies to meet urgent demands. Adaptations, in turn, may guide the development of more nuanced and effective food security policies in the future.

5 *Policy-program reuse.* Policy reuse refers to learning from one's own experiences, and differs from policy transfer—which involves learning from the experiences of others—different jurisdictions, governments, local governments, and bureaucracies (Dolowitz & Marsh, 2000; Rose, 1991). This distinction is crucial to understand learning processes under conditions of crisis and turbulence where uncertainties outweigh certainties and there is little time to learn from others. Policy reuse underscores the importance of leveraging current experiences to enhance preparedness, responsiveness, and effectiveness in future policymaking and program implementation. While policy diffusion is driven by factors such as competition, imitation, or mandates, leading to the spreading of ideas and practices, policy reuse is driven by efficiency and the desire to leverage existing work, focusing more on the practical benefits of replication. Typically, policy reuse involves taking an existing policy and applying it again within the same or a new context, often with specific adaptations. This empirical concept can be applied in two distinct food security scenarios. The first is where a local government implements an emergency food distribution network to ensure that vulnerable populations can access food despite lockdowns and supply chain disruptions. After the crisis, the city analyzes the effectiveness of the emergency food distribution network: in doing so, it identifies key strengths (such as rapid mobilization of volunteers) and weaknesses (such as logistical bottlenecks). By defining and mapping program reuse, there can be a better understanding of how new insights or successful contingent learning are integrated into future actions. These insights are then used to redesign the city's standard food aid programs, incorporating successful strategies from the crisis response. For example, the city may permanently adopt digital platforms for volunteer coordination and real-time monitoring of food deliveries, which proved effective during the pandemic. In a second scenario, a food security crisis in an urban area, which puts low-income households at risk, is addressed through crisis subsidies. Local authorities evaluate how well their emergency food programs performed and what could have been done better. However, lessons learned—such as the importance of stakeholder engagement, new services, or local food production and

diversified supply chains—are not applied to enhance ongoing food security programs.

6 *Return to routine*. The 'return to routine' mechanism refers to the process whereby innovative solutions developed during a crisis are institutionalized owing to their effectiveness, potential for replicability, and jurisdictional applicability. This process involves evaluating whether decisions and solutions, which may have been developed under emergency conditions, are likely to become stable enough to be integrated into regular policy frameworks. The effectiveness of a creative solution is tested over time to determine if it ought to remain beyond the crisis and become a standard part of the policy toolkit.

New policy instruments may initially seem adequate during a crisis due to the urgency and focus on immediate needs. However, the challenge is to assess whether these solutions remain effective once the crisis subsides and whether they ought to be institutionalized as part of the standard policy repertoire. In food security, this means determining whether the innovations developed during a food crisis continue to address the issue effectively and can be integrated into ongoing food safety or social policies. One may distinguish different scenarios of returning to the normal routine. The first scenario is *sustained innovation*: a new food assistance program, initially created as an emergency response, is evaluated and found to be highly effective. It is then institutionalized as a permanent part of the city's food security strategy. For example, a temporary food voucher system introduced during a crisis is adopted permanently, becoming a standard tool in a city's food security policy framework. The second scenario is *adaptation and integration*: an innovative solution developed during the crisis is adapted to fit into existing policy frameworks. It undergoes modifications based on post-crisis evaluations and is integrated into routine food security operations. For example, a community-based food distribution model used during a crisis is adapted to fit into a city's regular food assistance programs, improving its efficiency and reach. Finally, the third scenario is *re-evaluation and replacement:* it occurs when some innovative solutions may not prove effective or sustainable in the long term. They are re-evaluated and either replaced with new approaches or integrated into a different aspect of food security policy. It occurs when an unsustainable emergency food delivery system is replaced with a more viable long-term solution or its components are integrated into other policy areas, such as nutritional education, social security assistance, or local food production.

The following chapters examine governance mechanisms at the European, national, and urban levels, using the analytical framework outlined above. The research questions are: does poly-crisis shape the governance and street-level management of food aid programs? How do crises influence their ability to adapt and learn to effectively meet the needs of vulnerable populations? The empirical work to answer the research questions employs a range of techniques, each selected to address specific objectives within the analysis. These techniques include institutional documentary analysis, content analysis, in-depth interviews, and a comparative opposite-case analysis.

Institutional documentary analysis involves a detailed examination of laws, administrative acts, and institutional documents to enable an in-depth understanding of the formal policies, regulations, and strategies adopted at the European, national, and urban levels during crises. By reviewing these documents, the research uncovers the formal structures and legal underpinnings that shape food security governance and its evolution in response to crises, the role and resources assigned to street-level operators to interact with their beneficiaries.

Textual analysis is employed to systematically analyze European best practices and the key food security themes and narratives within the multilevel learning processes from the street level. This method is particularly useful to identify patterns in discourse, examining how food security issues are designed and addressed across different levels of governance. It helps to highlight the alignment in how food security is approached during crises at the street level in various contexts, from the European Union to individual cities.

In-depth interviews with local government officials, professionals, street-level bureaucrats, and other stakeholders provide rich qualitative data that offers insights into the mechanisms driving practical cases of food security governance at the urban level. These interviews are essential to understand how policy is implemented on the ground, the challenges faced by those working directly with food aid programs, and the collaborative learning processes that emerge in response to crises. The interviews also capture the perspectives of those directly involved in food security initiatives, providing a deeper understanding of the local adaptation strategies and the role of trust and collaboration in ensuring effective program delivery during crises.

Lastly, the longitudinal comparative analysis between two European cities as two opposite cases, such as the decentralized urban institutional context and network resource availability in Milan and the more centralized and scarce resource network availability in Warsaw, enable a detailed comparison in different learning processes and level of institutionalization in food security governance during crises. This comparison enables the identification of key differences and similarities in collaborative learning mechanisms, and the overall effectiveness of governance change in contrasting contexts. The use of documentary analysis, textual analysis, interviews, and comparative case studies provides a comprehensive understanding of the mechanisms that shape multilevel food security programs. This approach also highlights the role collaborative learning at street level plays in enhancing the adaptability and resilience of these programs during turbulent times.

Note

1. There are various definitions of discretion (Evans & Hupe, 2020). However, all disciplinary backgrounds concur that discretion entails some freedom granted to SLBs to make decisions they deem appropriate for the specific circumstances at hand (e.g., Hupe & Hill, 2007; Lipsky, 1980; Thomann et al., 2018; Tummers & Bekkers, 2014).

References

Aaberge, R., Brandolini, A., & Kyzyma, I. (2021). Threshold sensitivity of income poverty measures for EU social targets. In *Improving the understanding of poverty and social exclusion in Europe* (pp. 145–162). Publications Office of the European Union.

Abramson, M., & Mittu, R. (2006). Multi-agent coordination in open environments. In *Coordination of large-scale multiagent systems* (pp. 217–229). Springer US.

Action Aid. (2022). *Cresciuti troppo in fretta*. Action Aid, Milano.

Action Aid. (2023). *Frammenti da ricomporre*. Action Aid, Milano.

Agranoff, R. (2006). Inside collaborative networks: Ten lessons for public managers. *Public Administration Review, 66*(s1), 56–65. https://doi.org/10.1111/j.1540-6210.2006.00666.x

Ahn, S., & Norwood, F. B. (2020). Measuring food insecurity during the Covid-19 pandemic of Spring 2020. *Applied Economic Perspectives and Policy, 43*, 162–168.

Aizer, A. (2007). Public health insurance, program take-up, and child health. *The Review of Economics and Statistics, 89*(3), 400–415.

Allen, C. R., Fontaine, J. J., Pope, K. L., & Garmestani, A. S. (2011). Adaptive management for a turbulent future. *Journal of Environmental Management, 92*(5), 1339–1345.

Ansell, C. (2012). Collaborative governance. In *The Oxford handbook of governance*. https://doi.org/10.1093/oxfordhb/9780199560530.013.0035.

Ansell, C., Doberstein, C., Henderson, H., Siddiki, S. and 't Hart, P. (2020), Understanding inclusion in collaborative governance: A mixed methods approach, Policy and Society, 39(4), 570–591, https://doi.org/10.1080/14494035.2020.1785726.

Ansell, C., & Gash, A. (2008). Collaborative governance in theory and practice. *Journal of Public Administration Research and Theory, 18*(4), 543–571.

Ansell, C., & Gash, A. (2012). Stewards, mediators, and catalysts: Toward a model of collaborative leadership1. *The Innovation Journal, 17*(1), 2.

Ansell, C., Sørensen, E., & Torfing, J. (2021). The COVID-19 pandemic as a game changer for public administration and leadership? The need for robust governance responses to turbulent problems. *Public Management Review, 23*(7), 949–960.

Ansell, C., & Torfing, J. (2021a). Co-creation: The new kid on the block in public governance. *Policy & Politics, 49*(2), 211–230.

Ansell, C., & Torfing, J. (2021b). *Public governance as co-creation: A strategy for revitalizing the public sector and rejuvenating democracy*. Cambridge University Press.

Ansell, C. K., Trondal, J., & Øgård, M. (2017). *Governance in turbulent times*. Oxford University Press. Retrieved December 31, 2023, from https://books.google.com/books?hl=it&lr=&id=gwuDDQAAQBAJ&oi=fnd&pg=PP1&dq=ansell+etal+2017+Governance+in+turbulent+times&ots=QtpPEcAEeA&sig=ob9NH46e5oH_9Sbx_j39y3V37ig

Ansell, C. K., & Vogel, D. (Eds.). (2006). *What's the beef? The contested governance of European food safety*. MIT press.

Arnold, G. (2021). Distinguishing the street-level policy entrepreneur. *Public Administration, 99*(3), 439–453.

Aumann, R. J., & Maschler, M. (1964). The bargaining set for cooperative games. *Advances in Game Theory, 52*(1), 443–476.

Auqui-Caceres, M. V., & Furlan, A. (2023). Revitalizing double-loop learning in organizational contexts: A systematic review and research agenda. *European Management Review, 20*(4), 741–761.

Aviv, I., Gal, J., & Weiss-Gal, I. (2021). Social workers as street-level policy entrepreneurs. *Public Administration, 99*(3), 454–468.

Barrett, S. M. (2004). Implementation studies: Time for a revival? Personal reflections on 20 years of implementation studies. *Public Administration, 82*(2), 249–262. https://doi.org/10.1111/j.0033-3298.2004.00393.x

Barzelay, M. (2019). Public management as a design-oriented professional discipline. In *Public management as a design-oriented professional discipline*. Edward Elgar Publishing. Retrieved April 15, 2024, from https://www.elgaronline.com/monobook-oa/9781788119092/9781788119092.xml.

Bazzan, G. (2021). *Effective governance designs of food safety regulation in the EU: Do rules make the difference?* Springer Nature.

Bennett, C. J., & Howlett, M. (1992). The lessons of learning: Reconciling theories of policy learning and policy change. *Policy Sciences, 25*(3), 275–294.

Bergen, A., & While, A. (2005). 'Implementation deficit'and 'street-level bureaucracy': Policy, practice and change in the development of community nursing issues. *Health & Social Care in the Community, 13*(1), 1–10.

Bertelli, A. (2006). The role of political ideology in the structural design of new governance agencies. *Public Administration Review, 66*(4), 583–595.

Bertelli, A. M. (2021). *Democracy administered: How public administration shapes representative government.* Cambridge University Press.

Beutel, D., Tangen, D., & Carrington, S. (2023). Backward mapping from outcomes: Using a realist evaluation lens to evaluate an international aid program. *Evaluation and Program Planning, 97*, 102201.

Birkland, T. A. (2006). *Lessons of disaster: Policy change after catastrophic events.* Georgetown University Press.

Boin, A., Ekengren, M., & Rhinard, M. (2021). *Understanding the creeping crisis.* Springer Nature. https://doi.org/10.1007/978-3-030-70692-0

Boin, A., Hart, P. T., & Kuipers, S. (2018). The crisis approach. In *Handbook of disaster research* (pp. 23–38). Springer.

Boin, A., & Lodge, M. (2016). Designing resilient institutions for transboundary crisis management: A time for public administration. *Public Administration, 94*(2), 289–298.

Boin, A., & Rhinard, M. (2008). Managing transboundary crises: What role for the European Union? *International Studies Review, 10*(1), 1–26.

Brockmann, J. (2017). Unbureaucratic behavior among street-level bureaucrats: The case of the German state police. *Review of Public Personnel Administration, 37*(4), 430–451. https://doi.org/10.1177/0734371X15593990

Brodkin, E. Z. (2012). *Reflections on street-level bureaucracy: Past, present, and future.* Wiley Online Library.

Brodkin, E. Z. (2021). Street-level organizations at the front lines of crises. *Journal of Comparative Policy Analysis: Research and Practice, 23*(1), 16–29. https://doi.org/10.1080/13876988.2020.1848352

Browne, A., & Wildavsky, A. (1999). Implementation as mutual adaptation (1983). *Jeffrey L. Pressman and Aaron Wildavsky, Implementation* (3rd. ed.) (Berkeley: Univ. of California Press, 1984), 206–231.

Busetti, S., & Dente, B. (2018). *EXPOst. Le Conseguenze Di Un Grande Evento.* Retrieved December 31, 2023, from https://research.unite.it/handle/11575/104952.

Busetti, S., & Righettini, M. S. (2023). Policy learning from crises: Lessons learned from the Italian food stamp programme. *Policy & Politics, 51*(1), 91–112.

Cable, J., Jaykus, L. A., Hoelzer, K., Newton, J., & Torero, M. (2021). The impact of COVID-19 on food systems, safety, and security—A symposium report. *Annals of the New York Academy of Sciences, 1484*(1), 3–8.

Cafiero, C., Viviani, S., & Nord, M. (2018). Food security measurement in a global context: The food insecurity experience scale. *Measurement, 116*, 146–152.

Cairney, P., & Wellstead, A. (2021). COVID-19: Effective policymaking depends on trust in experts, politicians, and the public. *Policy Design and Practice, 4*(1), 1–14.

Calori, A., & Magarini, A. (2021). *The role of informal food networks. Notes from the Experience of Milan toward Food Policy Councils, Economia e Sostenibilità.*

Campomori, F., & Bassoli, M. (2024). Unravelling the multi-level governance of Ukrainian refugee reception in Italy: A focus on small and medium-sized cities. *Local Government Studies.* Retrieved August 30, 2024, from https://iris.unive.it/handle/10278/5067581.

Candel, J., & Daugbjerg, C. (2020). Overcoming the dependent variable problem in studying food policy. *Food Security, 12*(1), 169–178.

Candel, J. J. (2014). Food security governance: A systematic literature review. *Food Security, 6*, 585–601.

Candel, J. J. L., Breeman, G. E., & Termeer, C. J. A. M. (2016). The European Commission's ability to deal with wicked problems: An in-depth case study of the governance of food security. *Journal of European Public Policy, 23*(6), 789–813. https://doi.org/10.1080/13501763.2015.1068836

Capano, G., & Malandrino, A. (2022). Mapping the use of knowledge in policymaking: Barriers and facilitators from a subjectivist perspective (1990–2020). *Policy Sciences, 55*(3), 399–428.

Capano, G., Howlett, M., & Ramesh, M. (2015). *Varieties of governance: Dynamics, strategies, capacities.* Springer.

Capano, G., Howlett, M., & Ramesh, M. (2019). Disentangling the mechanistic chain for better policy design. In *Making policies work* (pp. 2–13). Edward Elgar Publishing. Retrieved June 14, 2023, from https://www.elgaronline.com/display/edcoll/9781788118187/9781788118187.00008.xml

Carless, D. (2019). Feedback loops and the longer-term: Towards feedback spirals. *Assessment & Evaluation in Higher Education, 44*(5), 705–714. https://doi.org/10.1080/02602938.2018.1531108

Carlson, S. J., Andrews, M. S., & Bickel, G. W. (1999). Measuring food insecurity and hunger in the United States: Development of a national benchmark measure and prevalence estimates. *The Journal of Nutrition, 129*(2), 510S–516S.

Charbonneau, B., & Giguère, A. (2025). The polycrisis and the uncertainty possibility space. *Global Sustainability, 8*, e4.

Chiwona-Karltun, L., Amuakwa-Mensah, F., Wamala-Larsson, C., Amuakwa-Mensah, S., Abu Hatab, A., Made, N., Taremwa, N. K., et al. (2021). COVID-19: From health crises to food security anxiety and policy implications. *Ambio, 50*(4), 794–811.

Cohen, N. (2021). *Policy entrepreneurship at the street level: Understanding the effect of the individual.* Cambridge University Press.

Cohen, N. (2022). An exploratory examination of the relationship between trust and street-level bureaucrats' willingness to risk their lives for others. *The American Review of Public Administration, 52*(3), 221–234.

Cohen, N., & Aviram, N. F. (2021). Street-level bureaucrats and policy entrepreneurship: When implementers challenge policy design. *Public Administration, 99*(3), 427–438.

Cohen, N., & Hertz, U. (2020). Street-level bureaucrats' social value orientation on and off duty. *Public Administration Review, 80*(3), 442–453.

Cohen, N., Lotta, G., Alcadipani, R., & Lazebnik, T. (2024). Trust and street-level bureaucrats' willingness to risk their lives for others: The case of Brazilian law enforcement. *The American Review of Public Administration, 54*(2), 119–134.

Cucciniello, M., & Linos, E. (2022). Institutional theory in behavioral public administration: A three-stage approach. In: The (missing?) role of institutions in behavioral public administration: A roundtable discourse. *Journal of Behavioral Public Administration, 5*(1), 9–13.

Davidovitz, M., & Cohen, N. (2022a). Playing defence: The impact of trust on the coping mechanisms of street-level bureaucrats. *Public Management Review, 24*(2), 279–300.

Davidovitz, M., & Cohen, N. (2022b). Which clients inspire or reduce the trust of street-level bureaucrats? *Administration & Society, 54*(8), 1516–1541.

Davidovitz, M., & Cohen, N. (2023). "I have learned my lesson": How clients' trust betrayals shape the future ways in which street-level bureaucrats cope with their clients. *Public Administration, 101*(1), 335–351.

Davidovitz, M., Cohen, N., & Gofen, A. (2021). Governmental response to crises and its implications for street-level implementation: Policy ambiguity, risk, and discretion during the COVID-19 pandemic. *Journal of Comparative Policy Analysis: Research and Practice, 23*(1), 120–130.

Dekker, S., & Hansén, D. (2004). Learning under pressure: The effects of politicization on organizational learning in public bureaucracies. *Journal of Public Administration Research and Theory, 14*(2), 211–230.

Deutsch, K. W. (1963). *The nerves of government: Models of political communication and control.* Retrieved December 2, 2024, from https://cir.nii.ac.jp/crid/1130282268630712960.

Deverell, E. (2009). Crises as learning triggers: Exploring a conceptual framework of crisis-induced learning. *Journal of Contingencies and Crisis management, 17*(3), 179–188.

Devine, D. (2024). Political trust and redistribution preferences. *Journal of European Public Policy,* 1–24. https://doi.org/10.1080/13501763.2024.2413194

Dewey, J., & Rogers, M. L. (2012). *The public and its problems: An essay in political inquiry*. Penn State Press.
Diaz-Granados, J., Sheehy, B., & Chen, Y. (2023). Rethinking global distributive justice: Legal and economic norms addressing crises of global health, hunger, and sustainability. https://papers.ssrn.com/sol3/papers.cfm?abstract_id=4878129#:~:text=http%3A//dx.doi.org/10.2139/ssrn.4878129.
Dickinson, H., & Sullivan, H. (2014). Towards a general theory of collaborative performance: The importance of efficacy and agency. *Public Administration, 92*(1), 161–177.
Dillenbourg, P. (1999). What do you mean by collaborative learning? In *Collaborative-learning: Cognitive and computational approaches* (pp. 1–19). Elsevier.
Dolowitz, D. P., & Marsh, D. (2000). Learning from abroad: The role of policy transfer in contemporary policy-making. *Governance, 13*(1), 5–23.
Dowler, E. A., & O'Connor, D. (2012). Rights-based approaches to addressing food poverty and food insecurity in Ireland and UK. *Social Science & Medicine, 74*(1), 44–51.
Durose, C. (2011). Revisiting Lipsky: Front-line work in UK local governance. *Political Studies, 59*(4), 978–995.
Eckersley, P., & Lakoma, K. (2022). Straddling multiple streams: Focusing events, policy entrepreneurs and problem brokers in the governance of English fire and rescue services. *Policy Studies, 43*(5), 1001–1020. https://doi.org/10.1080/01442872.2021.1892620
Edelenbos, J., & Klijn, E. H. (2007). Trust in complex decision-making networks: A theoretical and empirical exploration. *Administration & Society, 39*(1), 25–50. https://doi.org/10.1177/0095399706294460
Edge, S., & Meyer, S. B. (2019). Pursuing dignified food security through novel collaborative governance initiatives: Perceived benefits, tensions and lessons learned. *Social Science & Medicine, 232*, 77–85.
Ellis, S., & Shpielberg, N. (2003). Organizational learning mechanisms and managers' perceived uncertainty. *Human Relations, 56*(10), 1233–1254. https://doi.org/10.1177/00187267035610004
Elmore, R. F. (1979). Backward mapping: Implementation research and policy decisions. *Political Science Quarterly, 94*(4), 601–616.
Elston, T. (2017). Conflict between explicit and tacit public service bargains in U.K. executive agencies. *Governance, 30*(1), 85–104. https://doi.org/10.1111/gove.12191
Emerson, K., & Gerlak, A. K. (2016). Teaching collaborative governance online: Aligning collaborative instruction with online learning platforms. *Journal of Public Affairs Education, 22*(3), 327–344.
Emerson, K., & Nabatchi, T. (2015). *Collaborative governance regimes*, Georgetown University Press.

Emerson, K., Nabatchi, T., & Balogh, S. (2012). An integrative framework for collaborative governance. *Journal of Public Administration Research and Theory, 22*(1), 1–29.

Esobi, I. C., Lasode, M. K., Anyanwu, C. I., Barriguete, M. F., Okorie, M. A., & Lasode, D. O. (2021). Food insecurity, social vulnerability, and the impact of COVID-19 on population dependent on public assistance/SNAP: A case study of South Carolina, USA. *Journal of Food Security, 9*(1), 8–18.

Evans, T., & Hupe, P. L. (2020). *Discretion and the quest for controlled freedom*. Springer.

FAO's Agricolture and Development Economic Division (ESA). (2006). *Food Security*. Policy Brief, with support from the FAO Partnership Programme (FNPP) and the EC-GAO Foos Security Program.

Ferrario, M. N. (2014, April). The impacts on family consumption of the Bolsa Família subsidy programme. *CEPAL Review, 112*, 148–163.

Gerlak, A. K., & Heikkila, T. (2011). Building a theory of learning in collaboratives: Evidence from the everglades restoration program. *Journal of Public Administration Research and Theory, 21*(4), 619–644.

Gerlak, A. K., & Heikkila, T. (2019). *Tackling key challenges around learning in environmental governance*. Taylor & Francis.

Gerlack A. K., Heikkila T. & Lubell M., (2013). The promise and performance of collaborative governance. In *The Oxford handbook of US environmental policy*. Oxford University Press.

Gilardi, F., & Radaelli, C. M. (2012). Governance and learning. In D. Levi-Faure (Ed.), *The Oxford handbook of governance* (pp. 155–168). Oxford University Press.

Gofen, A., & Lotta, G. (2021). Street-level bureaucrats at the forefront of pandemic response: A comparative perspective. *Journal of Comparative Policy Analysis: Research and Practice, 23*(1), 3–15. https://doi.org/10.1080/13876988.2020.1861421

Gofen, A., Sella, S., & Gassner, D. (2019). Levels of analysis in street-level bureaucracy research. In *Research handbook on street-level bureaucracy*. Edward Elgar Publishing.

Greenwood, S., Singer, L., & Willis, W. (2021). *Collaborative governance: Principles, processes, and practical tools*. Routledge.

Gross, R., Schoeneberger, H., Pfeifer, H., & Preuss, H. J. (2000). The four dimensions of food and nutrition security: Definitions and concepts. *SCN News, 20*, 20–25.

Gulati, R., & Singh, H. (1998). The architecture of cooperation: Managing coordination costs and appropriation concerns in strategic alliances. *Administrative Science Quarterly, 43*, 781–814.

Hedström, P., & Swedberg, R. (1998). Social mechanisms: An introductory essay. In *Social mechanisms: An analytical approach to social theory* (pp. 1–31).

Retrieved December 3, 2024, from https://www.cambridge.org/core/services/aop-cambridge-core/content/view/1D9A78C32A006C6238BC55CA8DE58B9D

Hedström, P., & Ylikoski, P. (2010). Causal mechanisms in the social sciences. *Annual Review of Sociology, 36*(1), 49–67. https://doi.org/10.1146/annurev.soc.012809.102632

Heikkila, T., & Gerlak, A. K. (2013). Building a conceptual approach to collective learning: Lessons for public policy scholars. *Policy Studies Journal, 41*(3), 484–512.

Henderson, A. C. (2014). The critical role of street-level bureaucrats in disaster and crisis response. In *Handbook of critical incident analysis* (pp. 210–245). Routledge.

Henderson, J., Ward, P. R., Tonkin, E., Meyer, S. B., Pillen, H., McCullum, D., Toson, B., Webb, T., Coveney, J., & Wilson, A. (2020). Developing and maintaining public trust during and post-COVID-19: Can we apply a model developed for responding to food scares? *Frontiers in Public Health, 8*, 369.

Hough, M., Jackson, J., & Bradford, B. (2013). Legitimacy, trust and compliance: An empirical test of procedural justice theory using the European social survey. In J. Tankebe & A. Liebling (Eds.), *Legitimacy and criminal justice: An international exploration* (pp. 326–352). Oxford University Press.

Howlett, M. (2020). Dealing with the dark side of policy-making: Managing behavioural risk and volatility in policy designs. *Journal of Comparative Policy Analysis: Research and Practice, 22*(6), 612–625.

Howlett, M., & Ramesh, M. (2016). A chilles' heels of governance: Critical capacity deficits and their role in governance failures. *Regulation & Governance, 10*(4), 301–313.

Hupe, P., & Hill, M. (2007). Street-level bureaucracy and public accountability. *Public Administration, 85*(2), 279–299.

Hupe, P., & Hill, M. (2019). Positioning street-level bureaucracy research. In *Research handbook on street-level bureaucracy* (pp. 15–30). Edward Elgar Publishing.

Janssen, M., Chang, B. P. I., Hristov, H., Pravst, I., Profeta, A., & Millard, J. (2021). Changes in food consumption during the COVID-19 pandemic: Analysis of Consumer survey data from the first lockdown period in Denmark, Germany, and Slovenia. *Frontiers in Nutrition, 8*, 635859. Retrieved June 14, 2024, from https://www.frontiersin.org/articles/10.3389/fnut.2021.635859/full

Kamkhaji, J. C., & Radaelli, C. M. (2017). Crisis, learning and policy change in the European Union. *Journal of European Public Policy, 24*(5), 714–734.

Kjærnes, U., Dulsrud, A., & Poppe, C. (2006). Contestation over food safety: The significance of consumer trust. In C. Ansell & D. Vogel (Eds.), *What's the beef? The contested governance of European food safety* (p. 61). MIT Press.

Kleven, H. J., & Kopczuk, W. (2011). Transfer program complexity and the take-up of social benefits. *American Economic Journal: Economic Policy, 3*(1), 54–90.
Klijn, E. H., Edelenbos, J., & Steijn, B. (2010). Trust in governance networks: Its impacts on outcomes. *Administration & Society, 42*(2), 193–221. https://doi.org/10.1177/0095399710362716
Klijn, E. H., & Koppenjan, J. (2016). The impact of contract characteristics on the performance of public–private partnerships (PPPs). *Public Money & Management, 36*(6), 455–462.
Knight, L. (2002). Network learning: Exploring learning by interorganizational networks. *Human Relations, 55*(4), 427–454. https://doi.org/10.1177/0018726702554003
Lacchei, A. (2024). At the proof of time pressure: Asylum judges at the frontline of asylum appeals in Italy and France. *Rivista Italiana di Politiche Pubbliche, Rivista quadrimestrale, 2*, 219–244. https://doi.org/10.1483/114307
Lang, A., & Brüesch, C. (2018). Collaborative governance and co-production: The development of the e-relocation-notification in the Canton Zurich in Switzerland. In *XXII IRSPM Annual Conference*, Edinburgh, Scotland, 11–13 April 2018. Retrieved July 2, 2024, from https://digitalcollection.zhaw.ch/handle/11475/15735
Lang, A., & Brüesch, C. (2020). Collaborative governance in program implementation: The development of e-relocation notification in the Swiss Canton of Zurich. *International Journal of Public Administration, 43*(12), 1083–1095. https://doi.org/10.1080/01900692.2019.1665681
Lang, R., & Stoeger, H. (2018). The role of the local institutional context in understanding collaborative housing models: Empirical evidence from Austria. *International Journal of Housing Policy, 18*(1), 35–54. https://doi.org/10.1080/19491247.2016.1265265
Lelieveldt, H. (2023). Food industry influence in collaborative governance: The case of the Dutch prevention agreement on overweight. *Food Policy, 114*, 102380.
Levitt, B., & March, J. G. (1988). Organizational learning. *Annual Review of Sociology, 14*(1), 319–338.
Lilleker, D. G., & Stoeckle, T. (2021). The challenges of providing certainty in the face of wicked problems: Analysing the UK government's handling of the COVID-19 pandemic. *Journal of Public Affairs, 21*(4), e2733.
Linke, S., Erlwein, S., van Lierop, M., Fakirova, E., Pauleit, S., & Lang, W. (2022). Climate change adaption between governance and government—collaborative arrangements in the City of Munich. *Land, 11*(10), 1818. Retrieved July 2, 2024, from https://www.mdpi.com/2073-445X/11/10/1818
Lipsky, M. (1980). *Street-level bureaucracy: Dilemmas of the individual in public service*. Russel Sage Foundation.

Lizzi, R., & Righettini, M. S. (2022). Collaborative governance in Italian urban food policies: Towards an analytical framework for differentiated governance arrangements. *Rivista Italiana di Politiche Pubbliche, 17*(3), 301–317.
Lipsky, M., & Hill, M. (1993). Street-level bureaucracy: An introduction. In *The policy process: A reader* (pp. 381–385). Routledge.
Lodge, M. (2015). Accountability and consumer sovereignty. In *Accountability and regulatory governance* (pp. 235–264). Springer.
Lodge, M., & Wegrich, K. (Eds.). (2014). *The problem-solving capacity of the modern state: Governance challenges and administrative capacities.* Oxford University Press.
Lotta, G., Maynard-Moody, S., & Musheno, M. (2023). Practicing knowledge-agency in highly vulnerable communities: The frontlines of governance in delivering Brazilian human services. *Public Management Review*, 1–21. https://doi.org/10.1080/14719037.2023.2288250.
Lotta, G. S., & Marques, E. C. (2020). How social networks affect policy implementation: An analysis of street-level bureaucrats' performance regarding a health policy. *Social Policy & Administration, 54*(3), 345–360.
Ma, L., & Christensen, T. (2019). Government trust, social trust, and citizens' risk concerns: Evidence from crisis management in China. *Public Performance & Management Review, 42*(2), 383–404.
Maggetti, M., & Trein, P. (2019). Multilevel governance and problem-solving: Towards a dynamic theory of multilevel policy-making? *Public Administration, 97*(2), 355–369. https://doi.org/10.1111/padm.12573
Majone, G. (1989). *Evidence, argument, and persuasion in the policy process.* Yale University Press.
Malandrino, A., & Sager, F. (2021). Can teachers' discretion enhance the role of professionalism in times of crisis? A comparative policy analysis of distance teaching in Italy and Switzerland during the COVID-19 pandemic. *Journal of Comparative Policy Analysis: Research and Practice, 23*(1), 74–84. https://doi.org/10.1080/13876988.2020.1844544
Mark, S., Holder, S., Hoyer, D., Schoonover, R. & Aldrich, D. P. (2023, October 5). Understanding polycrisis: Definitions, applications, and responses. *Applications, and Responses.* https://doi.org/10.2139/ssrn.4593383
Mayer, R. C., Davis, J. H., & Schoorman, F. D. (1995). An integrative model of organizational trust. *Academy of Management Review, 20*(3), 709–734.
Maynard-Moody, S., & Musheno, M. (2000). State agent or citizen agent: Two narratives of discretion. *Journal of Public Administration Research and Theory, 10*(2), 329–358. https://doi.org/10.1093/oxfordjournals.jpart.a024272
Maynard-Moody, S., & Musheno, M. (2012). Social equities and inequities in practice: Street-level workers as agents and pragmatists. *Public Administration Review, 72*(s1), S16–S23.
Maynard-Moody, S. W., & Musheno, M. C. (2022). *Cops, teachers, counselors: Stories from the front lines of public service.* University of Michigan Press.

Mayne, Q., De Jong, J., & Fernandez-Monge, F. (2020). State capabilities for problem-oriented governance. *Perspectives on Public Management and Governance, 3*(1), 33–44.

McKeon, N. (2015). *Food security governance. Empowering communities, regulating corporations.* Routledge.

Mechlem, K. (2004). Food security and the right to food in the discourse of the United Nations. *European Law Journal, 10*(5), 631–648.

Melloni, E. (2024). Learning from place-based flood prevention policies: A realist approach. *Regional Studies*, 1–12. https://doi.org/10.1080/00343404.2024.2387140.

Meyer, R., & Kunreuther, H. (2017). *The ostrich paradox: Why we underprepare for disasters.* University of Pennsylvania Press.

Meyers, M. K., Glaser, B., & Donald, K. M. (1998). On the front lines of welfare delivery: Are workers implementing policy reforms? *Journal of Policy Analysis and Management: The Journal of the Association for Public Policy Analysis and Management, 17*(1), 1–22.

Meyers, M. K., & Nielsen, V. L. (2012). Street-level bureaucrats and the implementation of public policy. In *The SAGE handbook of public administration* (pp. 305–318). Sage.

Meza, O., Pérez-Chiqués, E., Campos, S. A., & Varela Castro, S. (2021). Against the COVID-19 pandemic: Analyzing role changes of healthcare street-level bureaucrats in Mexico. *Journal of Comparative Policy Analysis: Research and Practice, 23*(1), 109–119.

Migone, A., & Howlett, M. (2024). Assessing the "forgotten fundamental" in policy advisory systems research: Policy shops and the role(s) of core policy professionals. *Australian Journal of Public Administration, 83*(2), 192–214. https://doi.org/10.1111/1467-8500.12595

Millard, J., Sturla, A., Smutná, Z., Duží, B., Janssen, M., & Vávra, J. (2022). European food systems in a regional perspective: A comparative study of the effect of COVID-19 on households and city-region food systems. *Frontiers in Sustainable Food Systems*, 6. Retrieved October 20, 2023, from https://www.frontiersin.org/articles/10.3389/fsufs.2022.844170

Mishra, K., & Rampal, J. (2020). The COVID-19 pandemic and food insecurity: A viewpoint on India. *World Development, 135*, 105068.

Moloney, K., Jeong, B. G., & Sanabria-Pulido, P. (2025). New realities: Constancy of crisis and poly-crisis for public administration. In *The Routledge handbook on crisis, polycrisis, and public administration* (pp. 3–18). Routledge.

Moon, M. J. (2020). Fighting COVID-19 with agility, transparency, and participation: Wicked policy problems and new governance challenges. *Public Administration Review, 80*(4), 651–656.

Moseley, W. G., & Battersby, J. (2020). The vulnerability and resilience of African food systems, food security, and nutrition in the context of the COVID-19 pandemic. *African Studies Review, 63*(3), 449–461.

Moyson, S., Scholten, P., & Weible, C.M. (2017). Policy learning and policy change: Theorizing their relations from different perspectives. *Policy and Society, 36*(2), 161–77. Retrieved December 2, 2024, from https://academic.oup.com/policyandsociety/article-abstract/36/2/161/6403937

Newig, J., Challies, E., Jager, N. W., Kochskaemper, E., & Adzersen, A. (2018). The environmental performance of participatory and collaborative governance: A framework of causal mechanisms. *Policy Studies Journal, 46*(2), 269–297.

Nonaka, I., Von Krogh, G., & Voelpel, S. (2006). Organizational knowledge creation theory: Evolutionary paths and future advances. *Organization Studies, 27*(8), 1179–1208. https://doi.org/10.1177/0170840606066312

O'Connor, N., Farag, K., & Baines, R. (2016). What is food poverty? A conceptual framework. *British Food Journal, 118*(2), 429–449. https://doi.org/10.1108/BFJ-06-2015-0222

O'Hara, S., & Toussaint, E. C. (2021). Food access in crisis: Food security and COVID-19. *Ecological Economics, 180*, 106859. Retrieved June 14, 2024, from https://www.sciencedirect.com/science/article/pii/S0921800920312179

O'Toole, L. J., Jr., & Meier, K. J. (2006). Networking in the penumbra: Public management, cooptative links, and distributional consequences. *International Public Management Journal, 9*(3), 271–294. https://doi.org/10.1080/10967490600899580

O'Toole, L. J., Jr., Meier, K. J., & Nicholson-Crotty, S. (2005). Managing upward, downward and outward: Networks, hierarchical relationships and performance. *Public Management Review, 7*(1), 45–68.

OECD. (2024). *OECD survey on drivers of trust in public institutions—2024 results: Building trust in a complex policy environment.* OECD Publishing. https://doi.org/10.1787/9a20554b-en.

Ongaro, E. (2019). Advancing the theory and practice of public sector reform through the analysis of social mechanisms. In *Making policies work* (pp. 154–171). Edward Elgar Publishing.

Ortiz-Miranda, D., Moreno Pérez, O., & Arnalte Alegre, E. (2016). Food and nutrition security discursive frames in the context of the Spanish economic crisis. *Food Security, 8*(3), 665–677.

Ostrom, E. (2009). *Understanding institutional diversity.* Princeton University Press.

Ostrom, E. (2010). Beyond markets and states: Polycentric governance of complex economic systems. *American Economic Review, 100*(3), 641–672.

Ostrom, E., & Ahn, T. K. (2009). The meaning of social capital and its link to collective action. In *Handbook of social capital: The troika of sociology, political science and economics* (pp. 17–35). Edward Elgar.

Peters, B. G., & Pierre, J. (2000). Citizens versus the new public manager: The problem of mutual empowerment. *Administration & Society, 32*(1), 9–28.

Pleysier, S., & Kilkelly, U. (2024). Procedural justice and post-pandemic police-youth relations. *Youth Justice, 24*(2), 153–158.

Power, M., Doherty, B., Pybus, K. J., & Pickett, K. E. (2020). How COVID-19 has exposed inequalities in the UK food system: The case of UK food and poverty. *Emerald Open Research, 1*(10). https://doi.org/10.1108/EOR-10-2023-0003

Pressman, J. L., & Wildavsky, A. B. (1973). *How great expectations in Washington are dashed in Oakland*. University of California.

Puupponen, A., Huttunen, S., Kortetmäki, T., Lähteenmäki-Uutela, A., & Kaljonen, M. (2023). Justice in Finnish food policies. *Food Ethics, 8*(1), 6.

Radimer, K. L. (2002). Measurement of household food security in the USA and other industrialised countries. *Public Health Nutrition, 5*(6a), 859–864. https://doi.org/10.1079/PHN2002385

Radimer, K. L., Olson, C. M., Greene, J. C., Campbell, C. C., & Habicht, J. P. (1992). Understanding hunger and developing indicators to assess it in women and children. *Journal of Nutrition Education, 24*(1, Supplement 1), 36S–44S. https://doi.org/10.1016/S0022-3182(12)80137-3

Righettini, M. S., & Bazzan, G. (2017). From certification to coordination: Mechanisms for better results in European food safety policy. *Rivista Italiana Di Politiche Pubbliche, 12*(3), 309–332.

Righettini, M. S., & Lizzi, R. (2022). How scholars break down "policy coherence": The impact of sustainable development global agendas on academic literature. *Environmental Policy and Governance, 32*(2), 98–109. https://doi.org/10.1002/eet.1966

Rose, R. (1991). What Is lesson-drawing? *Journal of Public Policy, 11*(1), 3–30. Retrieved May 28, 2021, from https://www.jstor.org/stable/4007336.

Sager, F., & Gofen, A. (2022). The polity of implementation: Organizational and institutional arrangements in policy implementation. *Governance, 35*(2), 347–364.

Sager, F., Thomann, E., Zollinger, C., van der Heiden, N., & Mavrot, C. (2014). Street-level bureaucrats and new modes of governance: How conflicting roles affect the implementation of the Swiss ordinance on veterinary medicinal products. *Public Management Review, 16*(4), 481–502.

Sancino, A. (2022). *Public value co-creation: A multi-actor & multi-sector perspective*. Emerald Publishing Limited.

Sancino, A., Rees, J., & Schindele, I. (2018). Cross-sector collaboration for public value co-creation: A critical analysis. In *From austerity to abundance? Creative*

approaches to coordinating the common good (pp. 59–73). Emerald Publishing Limited.
Scerri, P., Vincent, R., & Mailler, R. (2006). Comparing three approaches to large-scale coordination. In *Coordination of large-scale multiagent systems* (pp. 53–71). Springer US.
Scharpf, F. W. (1997). Introduction: The problem-solving capacity of multi-level governance. *Journal of European Public Policy, 4*(4), 520–538.
Scognamiglio, F., Sancino, A., Caló, F., Jacklin-Jarvis, C., & Rees, J. (2022). The public sector and co-creation in turbulent times: A systematic literature review on robust governance in the COVID-19 emergency. *Public Administration, 101*, 53–70.
Scott, I., & Gong, T. (2021). Coordinating government silos: Challenges and opportunities. *Global Public Policy and Governance, 1*(1), 20–38.
Scott, T. A., & Craig, W. T. (2017). Unpacking the collaborative toolbox: Why and when do public managers choose collaborative governance strategies? *Policy Studies Journal, 45*(1), 191–214.
Silano, M., & Silano, V. (2020). *Ensuring food safety in the European Union*. CRC Press. Retrieved December 3, 2024, from https://www.taylorfrancis.com/books/mono/10.1201/9781003088493/ensuring-food-safety-european-union-marco-silano-vittorio-silano.
Simon H. A. (1947). *The administrative behavior*. Macmillan.
Simon, H. A. (2013). *Administrative behavior*. Simon and Schuster.
Smith, D., & Thompson, C. (2023). The complexities and challenges of researching hunger in wealthier countries. In *Food deserts and food insecurity in the UK* (pp. 1–7). Routledge.
Sorensen, C. H., & Gudmundsson, H. (2010). The impact of governance modes on sustainable transport–the case of bus transport in greater Manchester, UK. *World Review of Intermodal Transportation Research, 3*(1–2), 8–25.
Stensöta, H. O. (2011). Political influence on street-level bureaucratic outcome: Testing the interaction between bureaucratic ideology and local community political orientation. *Journal of Public Administration Research and Theory, 22*(3), 553–571.
Stirton, L., & Lodge, M. (2001). Transparency mechanisms: Building publicness into public services. *Journal of Law and Society, 28*(4), 471–489.
Tamuz, M. (2001). Learning disabilities for regulators: The perils of organizational learning in the air transportation industry. *Administration & Society, 33*(3), 276–302.
Taylor, K., DeLeo, R. A., Zarb, S., Jeschke, N., & Birkland, T. A. (2023). Subnational focusing events and agenda change: The case of toxic algae bloom and contaminated drinking water in Toledo, Ohio. In *A modern guide to the multiple streams framework* (pp. 222–245). Edward Elgar Publishing.

Thomann, E. (2015). Is output performance all about the resources? A fuzzy-set qualitative comparative analysis of street-level bureaucrats in Switzerland. *Public Administration, 93*(1), 177–194.

Thomann, E., van Engen, N., & Tummers, L. (2018). The necessity of discretion: A behavioral evaluation of bottom-up implementation theory. *Journal of Public Administration Research and Theory, 28*(4), 583–601.

Trein, P., Maggetti, M., & Meyer, I. (2021). Necessary conditions for policy integration and administrative coordination reforms: An exploratory analysis. *Journal of European Public Policy, 28*(9), 1410–1431.

Trein, P., Thomann, E., & Maggetti, M. (2019). Integration, functional differentiation and problem-solving in multilevel governance. *Public Administration, 97*(2), 339–354.

Tschersich, J., & Kok, K. P. (2022). Deepening democracy for the governance toward just transitions in Agri-food systems. *Environmental Innovation and Societal Transitions, 43*, 358–374.

Tummers, L., & Bekkers, V. (2014). Policy implementation, street-level bureaucracy, and the importance of discretion. *Public Management Review, 16*(4), 527–547.

Tyler, T. R. (1988). What is procedural justice?: Criteria used by citizens to assess the fairness of legal procedures. *Law & Society Review, 22*(1), 103–135.

Tyler, T. R. (2006). *Why people obey the law*. Princeton University Press.

Van Acker, W., & Bouckaert, G. (2018). What makes public sector innovations survive? An exploratory study of the influence of feedback, accountability and learning. *International Review of Administrative Sciences, 84*(2), 249–268. https://doi.org/10.1177/0020852317700481

Vangen, S., & Huxham, C. (2013). Building and using the theory of collaborative advantage. In R. Keast, M. P. Mandell, & R. Agranoff (Eds.), *Network theory in the public sector* (pp. 51–67). Routledge.

Verhoest, K., Redert, B., Maggetti, M., Levi-Faur, D., & Jordana, J. (2025). Trust and regulation. In *Handbook on trust in public governance* (pp. 360–380). Edward Elgar Publishing.

Vike, H. (2018). Street-level bureaucracy and crosscutting cleavages in municipal worlds. In *Bureaucracy and society in transition: Comparative perspectives* (pp. 245–263). Emerald Publishing Limited.

Walker, G. B., & Daniels, S. E. (2019). Collaboration in environmental conflict management and decision-making: Comparing best practices with insights from collaborative learning work. *Frontiers in Communication, 4*, 2.

Weiss, C. H. (1980). Knowledge creep and decision accretion. *Knowledge, 1*(3), 381–404. https://doi.org/10.1177/107554708000100303

Wild, C. E., Conceição, M. I. G., Iwakuma, M., Lewis-Jackson, S., Toyomoto, R., de Souza, A. R. N. D., & Rai, T. (2023). Perceptions of government guidance

and citizen responses during the COVID-19 pandemic: A cross-country analysis. *SSM-Qualitative Research in Health, 4*, 100308.
Windfuhr, M. (2010). The world food crisis and the right to adequate food. *Universal Human Rights and Extraterritorial Obligations, 130*, 148.
Windfuhr, M., & Jonsén, J. (2005). *Food sovereignty. Towards democracy in localized food systems.* ITDG Publishing.
Wiskerke, J. S. (2009). On places lost and places regained: Reflections on the alternative food geography and sustainable regional development. *International Planning Studies, 14*(4), 369–387.
Wiskerke, J. S., & Viljoen, A. (2012). Sustainable urban food provisioning: Challenges for scientists, policymakers, planners and designers. In *Sustainable food planning: Evolving theory and practice* (pp. 19–36).
Zeitlin, J. (2016). EU experimentalist governance in times of crisis. *West European Politics, 39*(5), 1073–1094. https://doi.org/10.1080/01402382.2016.1181873
Zhang, L., Zhao, J., & Dong, W. (2021). Street-level bureaucrats as policy entrepreneurs: Action strategies for flexible community governance in China. *Public Administration, 99*(3), 469–483.

CHAPTER 2

Food Insecurity in Europe: Governing Amid a Poly-crisis

The governance mechanisms of European food security evolved in the global poly-crisis context between 2014 and 2021. The financial and economic crisis had emphasized the fragility of food availability, access, and stability (Galanakis, 2020), revealing that more than mere pre-established management plans were needed to address new, evolving, and complex challenges (Mintzberg & McHugh, 1985). To enhance Member States' capacity to tackle rising social vulnerabilities, including food insecurity, the EU has redefined its approach through the Fund for European Aid to the Most Vulnerable (FEAD) and fostered a collaborative learning process among public and private actors involved in its management. This chapter examines the reframing of food insecurity at the European governance level and the collaborative multilevel learning process activated through the FEAD administration. The pre-pandemic food security challenges prompted the EU and Member State governments at multiple levels to adopt a more holistic and integrated approach to address social vulnerability. In such an approach, food security played a new central role. Increasingly formalized public-private partnerships with the third sector were boosted to increase the capacity to address needs in a bottom-up way and to build more resilient communities across Europe. Such alliances, particularly with non-governmental organizations (NGOs) and community groups, would later prove crucial in managing the emergency response during the pandemic crises (Ansell et al., 2017; Boyne & Meier, 2009).

M. S. Righettini, *The Governance of Food Security in Europe*,
https://doi.org/10.1007/978-3-031-98083-1_2

2.1 Shifting Focus: Rising Awareness of Food Vulnerability

The Food and Agriculture Organization's (FAO) reports since 2017 have consistently shown that at a global level, conflicts, extreme weather events, economic downturns, and pandemics are increasing conditions of unaffordable nutritious foods, and growing inequality is derailing efforts to meet the 2023 Agenda targets. Moreover, 'the relative lack of change in hunger at the global level from 2021 to 2022 due to the pandemic effects (see Fig. 2.1) hides substantial differences at the regional level, and it is still higher in low-income countries' (FAO, IFAD, UNICEF, WFP and WHO, 2023, pp. 13–15). According to the report entitled *The State of Food Security and Nutrition in the World (SOFI)* (FAO, IFAD, UNICEF, WFP, and WHO, 2019), between 2016 and 2018, an estimated 821 million people were undernourished, with the majority living in low-income

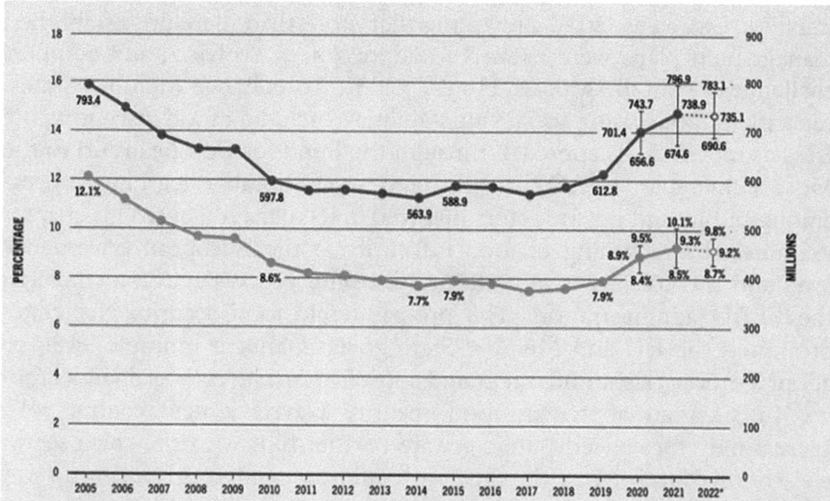

Fig. 2.1 Global hunger trend from 2005 to 2022. Pre- and post-COVID-19 levels. (Source: FAO, 2023, p. 8, https://openknowledge.fao.org/server/api/core/bitstreams/f1ee0c49-04e7-43df-9b83-6820f4f37ca9/content/state-food-security-and-nutrition-2023/food-security-nutrition-indicators.html)

countries, where 12.9% of the population suffered from undernourishment.

Before the COVID-19 pandemic, European institutions already had significant concerns about food security and nutrition (FSN) quality. Looking specifically at the data by country, there are variations in the prevalence of moderate or severe food insecurity as measured by the Food Insecurity Experience Scale (FIES). The indicator measuring the inability to afford a meal in the EU decreased by an average of 2.6 percentage points overall and by 4.1 percentage points for poorer households between 2010 and 2019 (EU, 2024). However, the COVID-19 pandemic, along with recent disruptions caused by Russia's aggression against Ukraine, halted and reversed much of the progress made over the past decade. The portion of the population facing moderate or severe constraints on their ability to obtain sufficient food over a year is not uniform: Italy, Germany, Ireland, and the United Kingdom have a food insecurity rate of 2.5%, while Bulgaria and Romania have approximately double this rate. The situation worsened owing to the COVID-19 pandemic, particularly in urban areas. Those classifies as 'most deprived' under the EU intervention were migrants, informal-economy workers, and people in precarious jobs— especially in services, hospitality, and retail — suffered major job losses, due to social distancing measures and the wider economic downturn.[2] The pandemic heavily affected the four pillars of the FAO's food security and nutrition (FSN) forum: availability, access, utilization, and stability. This disruption and growing inequalities further restricted vulnerable populations' ability to make informed and healthy decisions within food systems. According to FAOSTAT (cited July, 2023, p. 6), after the pandemic, severe food insecurity in Europe including the UK affects 2% of the population, while moderate and severe food insecurity affects 7% of the population overall. The COVID-19 pandemic, the energy crisis, and the inflation caused by the Russia-Ukraine war have increased the turbulence and the complexity of food security governance. Turbulence reduces the period of validity of consolidated public programs and creates opportunities for a game changer for governments in food assistance programs (Busetti & Righettini, 2023). New challenges are visible above all in urban contexts, where food system crises converge and intertwine, affecting people and households, and where public programs are called upon to address and improve the living conditions of the local communities (López Cifuentes & Fiala, 2022; Nordensvärd et al., 2023). In urban contexts, food security's *wickedness* is anything but an 'abstraction' or an 'empty' concept (Morin, 1992; Morin, 2017); it becomes the measure of the vulnerability in living conditions of citizens who ask for tangible solutions to specific food problems. Despite the increasing social salience on the issue, access to food and food poverty often does not find a specific definition and position at a government and local policy level. Food security is often viewed as a 'secondary' and 'collateral' declination of other programs and policy tools: income support, housing, poverty, immigration, and health problems. To fully grasp the novelties in European food security governance, one ought to distinguish between food security, food poverty, and food vulnerability. Food security is an ideal state where all people have stable access to available nutritious food, whereas food poverty is an observed condition of deprivation, and vulnerability is the risk of falling into food insecurity due to external shocks (FAO, 2023). Food poverty is a narrower, descriptive, and rights-based concept referring to the inability to afford or access an adequate diet due to low income or lack of resources (Dowler & O'Connor, 2012; O'Connor et al., 2016). Vulnerability refers to the degree to which individuals or groups are at risk of experiencing food insecurity owing to external shocks or systemic weaknesses (Ericksen, 2008; Esobi et al., 2021). During crises, the three concepts—food security, food poverty, and vulnerability—become deeply interconnected, forming a cause-effect chain. As vulnerabilities increase, food availability, access, utilization, and stability are affected, leading to a decline in food security. When a crisis reduces food security, food poverty becomes tangible for the affected households and individuals that face higher vulnerability to future crises.

Poly-crisis has increased food vulnerability, as well as salience and attention to food needs in Europe. In response, the adoption of Food Aid Programs (FAPs) by European, national, and local governments have become a pre- and post-COVID trend (Madama, 2018). The food assistance policy in Europe has a long history, even though its rationale has changed over time. The 1996 World Food Summit, organized by the Food and Agriculture Organization (FAO), underscored the importance of food security, influencing the EU to incorporate food security considerations into its external policies, particularly about development and humanitarian aid. This period marked a major evolution in the EU's approach to food security, recognizing it not only as a domestic agricultural concern but also as a key social factor in global stability and development. In recent years, food insecurity and disparities in access to food have become increasingly prominent social challenges within European policy debates. These challenges have been framed in ways that emphasize the need for comprehensive, cross-sectoral, and cross-country policy measures designed both to alleviate and prevent food insecurity—encompassing agricultural, fisheries, social, and economic sectors, which are insufficient and ineffective in addressing the complexities of food insecurity (Kumareswaran & Jayasinghe, 2022). The recognition of food insecurity's multidimensionality extends beyond European institutions and reaches national, regional, and local governance levels, where policies can have more direct and immediate impacts on communities in need. Localized approaches are seen as crucial for crafting tailored solutions that meet the specific needs of vulnerable populations, and they advocate for a governance approach that leverages the nexuses between the different dimensions of food security—such as nourishment, availability, access, utilization, and stability (Hermans et al., 2023; Hermans & Cantillon, 2025). Such an approach emphasizes multi-actor collaboration and governance mechanisms to boost consistency and horizontal integration, especially in urbanization and shifting socio-economic conditions. Recognizing the interplay between different food security dimensions and between food security, vulnerability and poverty aligns with broader efforts to adapt governance frameworks and tools to new contextual challenges, particularly in times of crisis (Moragues-Faus & Battersby, 2021; Nordensvärd et al., 2023).

The COVID-19 pandemic and the Russia-Ukraine war, combined with the impacts of climate change, are recognized as having profound economic and societal effects (Adamczak-Retecka & Śniadach, 2021); nevertheless, the epistemological challenges in the field of food security are longstanding ones and revolve around the complexities of defining, measuring, and addressing food security within diverse contexts. These challenges affect how knowledge is produced, validated, and applied to policymaking processes. In some cases, the production of food security knowledge may be influenced by political or ideological biases. For example, EU agricultural subsidies under the Common Agricultural Policy (CAP) are highly political. They can shape research priorities that may not align with the most pressing food security concerns, such as sustainability or social equity. Policymakers face epistemological challenges in reconciling competing agendas, which may prioritize different types of evidence (e.g., economic growth vs. social equity vs. environmental sustainability). This can lead to fragmented or inconsistent policy approaches, where some dimensions of food security are addressed while others are neglected. It is essential to understand the critical role that the definition, measurements, and monitoring of food security have played and are still playing in designing and redesigning food security governance, especially from a multilevel perspective (Pérez-Escamilla et al., 2017). The epistemological challenges affecting food security are reflected (Carlson et al., 1999) in many methods and techniques available to monitor food insecurity at a broader international and global as well as individual level. Different approaches reveal the complexity of navigating data reliability issues, causal complexity, political biases, and the tension between local and global scales. For example, the FAO's Food Insecurity Experience Scale (FIES) is widely used for national food access and hunger monitoring, complementing other hunger indicators such as the Prevalence of Undernourishment (SDG 2.1.1). It gathers data on household/individual experiences with food insecurity through eight yes/no questions, offering insights into the severity of food insecurity. While the FIES is useful in reflecting people's experiences and identifying vulnerable groups, it has some limitations in addressing the underlying causes of food insecurity (e.g., unemployment or supply chain issues).

The Global Food Security Index (GFSI), developed by the Economist Group, tracks food security through 58 indicators across four dimensions: affordability, availability, quality and safety, and sustainability. The GFSI shows a deterioration in global food security from 2020 to 2022 due to the COVID-19 pandemic and rising food prices. Affordability, the most crucial pillar, experienced a 4% decline during this period. Top-performing countries in the GFSI include Finland, Ireland, and Norway, while countries in Sub-Saharan Africa and conflict zones such as Syria and Haiti scored the lowest. The Integrated Food Security Phase Classification (IPC) offers a standardized framework for assessing the severity of acute and chronic food insecurity and malnutrition. It differentiates between various forms of food insecurity and malnutrition, helping policymakers design specific interventions. The IPC includes diverse stakeholders in analyzing food security, including governments, UN agencies, and NGOs. Public and private organizations emphasize the need for robust data collection at both national and local levels to ensure targeted and effective food security interventions. Action Aid developed the index of Material and/or Social Food Deprivation (Deprivazione Alimentare, Materiale e/o Sociale—DAMS) to measure hunger outcomes and government commitments to eradicating hunger, with a focus on legal rights to food and investments in agriculture (Action Aid, 2024).[1] Many measurement tools focus on food availability and affordability but fail to capture governance effectiveness, policy implementation, and systemic inequalities in food systems. Meso-level assessments (e.g., tracking policy interventions, evaluating government accountability) remain underdeveloped, limiting their ability to inform targeted policy interventions.

2.2 From Agriculture to Social Governance Salience: The Evolution of EU Food Aid Governance

The difficulty of seriously tackling food insecurity and aid in industrialized countries can be partially attributed to the challenge of comprehending the intricacies of the issue within a multilevel governance configuration and jurisdiction. In Europe, social inequality is generally recognized and addressed through national legislation, including food aid programs that are often integrated within the broader framework of social inequality policies. While a staunchly protected national sovereignty marks the domain of food poverty alleviation and social inclusion, the significance and effectiveness of food security strategies and tools are decided at street level. Where urban governments are responsible for implementing national policies and managing local welfare, non-state organizations, including charities and non-profits, often take the lead in assistance. The European Union first addressed food poverty between 1987 and 2013 within the Common Agricultural Policy (CAP) framework. This initiative, made possible by the availability and strategic use of food stocks, facilitated the distribution of hundreds of tons of food each year to those most in need. In 1987, Jacques Delors, then President of the European Commission, championed launching the European Programme for distributing Food to the Most Deprived (PEAD). PEAD provided crucial support to charities and civil society organizations, helping them ensure that essential goods reached the EU's most deprived citizens. Established during an exceptionally severe winter, the Programme initially enabled Member States to release surplus agricultural stocks, distributing them as food aid to needy people. Over time, PEAD evolved into a more comprehensive model, reaching an expanding number of beneficiaries, and increasing its impact. By 2011, the Programme assisted almost 19 million people (18,986,182), 35% of the estimated poor population across 20 EU Member States. PEAD functioned as a supplementary resource, enhancing the efforts of existing organizations dedicated to aiding vulnerable populations while reinforcing the capacity of traditional, locally rooted networks rather than establishing new, government-run supply chains. The dismantling of PEAD in 2013 was driven by various factors, including the CAP reform and the view held by some Member States that such a program, while beneficial, was better suited to social policy initiatives and national jurisdiction rather than being embedded within agricultural policy. For countries with robust income-support mechanisms, PEAD was seen as less aligned with their poverty reduction frameworks. However, with the need for essential assistance still pressing, the call emerged for a new EU-wide instrument that could supplement national policies and continue to address deprivation more sustainably. This approach aligned with the 'Europe 2020' strategy, which aimed to reduce the number of people facing poverty and social exclusion by 20 million across the EU by 2020. Following extensive impact assessments, discussions, and consultations, in March 2014, the European Parliament and the Council established the Fund for European Aid to the Most Deprived (FEAD), a dedicated funding program to tackle the growing social exclusion resulting from the global financial crisis.[2] The FEAD was established in continuity with the previous program (PEAD) to respond to the economic and social challenges that had emerged, particularly during the economic and multi-year debt crisis that took place in the Eurozone from 2009 until the mid to late 2010s and the Great Recession of 2008–2012 (Loopstra et al., 2015). The FEAD was a significant shift from prior approaches, from Member States' exclusive responsibility for addressing extreme poverty to a coordinated EU initiative. One of the FEAD's key innovations was that it expanded its scope beyond food aid. Its mission now included both food and essential material assistance, with Member States empowered to choose between two Operational Programmes (OPs) for the 2014–2020 period: *Operational Programme I* focused on 'food and/or basic material assistance,' while *Operational Programme II* aimed at 'the social inclusion of the most deprived.' This expanded mandate responded to requests from Member States preferring to allocate resources to broader social support measures rather than limiting them to food and material distribution. While the FEAD budget remained consistent with the PEAD's budget, the governance and focus underwent significant change. The Fund for European Aid to the Most Deprived (FEAD) administration has undergone a significant shift from an agricultural-focused approach to one centered on material assistance and social programs. The change reflects an evolution in the fund's purpose, moving from an initial aim of poverty reduction through surplus food distribution to a broader commitment to addressing food vulnerability and social exclusion. With this new approach, each Member State was tasked with developing Operational Programmes (OPs) tailored to specific social needs and priorities. The change in governance can be seen as a shift in the institutional agenda from an indirect food support strategy—focused on providing financial aid to producers and investing in infrastructure—to a direct approach involving monetary transfers and food supply to intermediate beneficiaries (third-sector organizations) and, ultimately, to final beneficiaries (those in need) (Alderman et al., 2018). The shift involved collaborating with various stakeholders to ensure the programs were comprehensive and tailored to existing networks. Stakeholders could include national, regional, and local authorities, public institutions, non-profit organizations, and other entities directly involved in the social welfare and food aid sectors. This inclusive approach ensures that the FEAD's resources are allocated effectively, reaching intermediate beneficiaries and vulnerable populations through coordinated efforts across all levels of governance. The formalized partnership-driven governance model underscores a new governance structure, increasing the number of actors involved and the governance mechanism—such as conditionality, horizontal and vertical coordination—in addressing social inclusion at the Member State level. In a further step, after the outbreak of the pandemic, the European Commission and the European Parliament approved a new financial tool, the European Social Fund Plus (ESF+), in 2022, which integrates the former European Social Fund (ESF), the Youth Employment Initiative, the Fund for European Aid to the Most Deprived, and the EU Programme for Employment and Social Innovation. Thus, some characteristics of food security governance were changed again (Table 2.1).

As European institutions navigated an increasingly complex poly-crisis landscape (Tashtoush et al., 2019), the need to re-evaluate and strengthen more inclusive monitoring practices as essential policy tools became urgent (Dinan et al., 2024). Appropriate indicators are essential to monitor societal trends during turbulent times, effectively address overlapping challenges, identify emerging risks, and assess the impact of policy interventions. Furthermore, indicators foster accountability and transparency by offering measurable benchmarks through which the success of policies could be evaluated. Monitoring indicators of food poverty and insecurity were becoming passive reflections of reality rather than active tools to shape governance mechanisms and policy priorities, guiding resource allocation, and to facilitate coordination across governance levels. The development

Table 2.1 EU food security governance transition: structure and mechanisms. From the PEAD to FEAD and ESF+ (1987–2027)

Governance scope	**PEAD (1987–2013)**	**FEAD (2014–2020–2021)**	**ESF+ (2021–2027)**
Focus and value	Food production and provision Food poverty mitigation through redistribution of surplus agricultural products to people in need	Social inclusion Food vulnerability reduction: a multidimensional approach to food insecurity linked to social vulnerability	Social rights Absolute poverty and material deprivation: Food vulnerability disappears from the new fund priority areas (migrants and children)
Governance structure	**PEAD (1987–2013)**	**FEAD (2014–2020–2021)**	**ESF+(2021–2027)**
Number of EU member states involved	On a voluntary basis	All the 28 member states	All the 27 member states
Other actors involved	Private organizations (third sector)	Public (national, regional, and local) and private organizations	Public (national, regional, and local) and private organizations
European and national sub system jurisdiction	CAP—Agriculture	Employment, social affairs, and inclusion	Employment, social affairs, and inclusion
Governance mechanisms	**PEAD**	**FEAD**	**ESF+ (21–27)**
Participation of states in the fund	Voluntary (20 MS opt-in)	Compulsory	Compulsory
Financial participation of the states in the FUND	Voluntary decision	Threshold-based decision. Compulsory at 15% of the eligible expenses	At least 3% of ESF+ must be allocated to food and basic material assistance for the needy
Type of assistance	Food products	Food products and/or other basic consumer goods and social inclusion measures	Adequate funding is also due for capacity building of social partners and civil society in member states, and at least 0.25% of the fund should be allocated based on country-specific recommendations

(*continued*)

Table 2.1 (continued)

Partnerships	Member states and food producers	Member states, third sectors and limited number of municipalities	Member states, third sectors, regions and municipalities
Type of coordination	Disjointed and not formalized	Vertical through European brokerage—And horizontal though formalized and integrated relations	Multilevel and multi-actor: National, regional, and local actions within the third sector
Accountability reports	None	Compulsory	Compulsory

Source: Made by the author based on Marsico and Frigo (2014)

of a new complex indicator occurred in Europe in 2010 when the European social monitoring system adopted the AROPE index (people at risk of poverty or social exclusion): this is a tool for monitoring the implementation of the Europe 2020 strategy based on a multidimensional concept of vulnerability and social exclusion including food insecurity (Chaudhuri, 2003; Moret, 2014). The AROPE index identifies vulnerable individuals residing in households with a total equivalized disposable income below a country-specific threshold, set at 60% of the national median equivalized disposable income (Guio et al., 2021, p. 26). The introduction of this comprehensive indicator marked a transformative shift toward a multilevel, more socially integrated European food security policy.

The new monitoring system placed food assistance within a holistic framework of social welfare and material support, connecting multiple and diverse local systems to unify and strengthen their efforts across communities (Jochim & May, 2010).

The adoption of a vulnerability index contributed to shaping multilevel networking, where local, regional, and national governments work with EU institutions. The collection, processing, and interpretation of data for the vulnerability index enabled better interinstitutional communication, policy agenda alignment and a more targeted, evidence-based resource distribution, ensuring that policies are better suited to the specific needs of different groups or areas. The *people at risk of poverty or social exclusion* indicator is the outcome of a growing institutional awareness that food is an integral part of poverty and social exclusion and, therefore, needs to be measured and addressed in Europe. The index provided a more nuanced understanding of food security and helped policymakers implement holistic views on problem coupling and inclusive solutions and interventions. This indicator also shaped multilevel governance approaches, where local, regional, and national governments work together with EU institutions. The AROPE index incorporates a cross-sectoral approach and captures various dimensions of vulnerability in a more holistic way: income poverty, food and water poverty (access to adequate food), energy poverty (adequate housing), health and educational poverty (Guio et al., 2021). Reflecting on the intersection between food and other types of vulnerability is a shift from an income-driven, unidimensional approach to a more comprehensive, multidimensional approach and intersectoral strategies for managing access to food and related resources. In fact, at the urban level, food insecurity is driven not just by the mere availability of food but also by the ability to access quality food, use resources effectively, and navigate social power dynamics. This approach acknowledges the importance of institutional actors in shaping and ensuring sustainable and stable access to food resources (Amartya Sen, 1982). The 'people at risk of poverty or social exclusion' indicator underscores the need for comprehensive measurement processes and targeted action across Europe. It is a valuable tool for policymakers, as it enables them to adopt more holistic perspectives that integrate interconnected challenges and design inclusive, sustainable solutions.

The AROPE rate monitors 'the share of the total population which is at risk of poverty or social exclusion' (Guio 2021), which 'corresponds to the sum of persons who are either at risk of poverty or severely materially and socially deprived or living in a household with a very low work intensity.' The *Severe Material and Social Deprivation* (SMSD) rate indicates the 'Proportion of the population experiencing an enforced lack of at least 7 out of 13 deprivation items' (6 related to the individual and 7 related to the household) including being able to afford a meal with meat, chicken, fish, or a vegetarian equivalent every second day.

The interplay between the Severe Material and Social Deprivation (SMSD) rate, the At-Risk-of-Poverty Rate, and the Low Work Intensity Indicator reveals the multifaceted nature of vulnerability and its impact on access to essential resources such as food. The Severe Material and Social Deprivation (SMSD) rate indicates the intensity of material hardship, including the inability to afford necessities such as regular meals. It highlights the link between food insecurity and broader deprivation, using criteria such as being able to afford a protein-rich or vegetarian equivalent meal every other day. The At-Risk-of-Poverty Rate reflects economic vulnerability by measuring disposable income levels below the national median, emphasizing the fact that a low income limits access to basic needs, including nutritious food. The Low Work Intensity Indicator explores the relationship between employment and food poverty, showing that households with limited work intensity often lack sufficient income to secure adequate food.

These indicators underscore the complex interconnection between economic hardship, employment patterns, and food insecurity as a fundamental element of social insecurity. The EU's integrated approach to food security involved a multifaceted strategy: allocating financial resources, collecting robust data, employing comprehensive statistics to inform policies, and fostering multilevel governance focusing on the active involvement of third-sector organizations and local governments.

2.3 The FEAD Network: Strengthening Social Cohesion in Times of Crisis

Under the FEAD program (2014–2020), Member States tailored initiatives to their needs through two operational programs: OP I, focusing on food and material aid, and OP II, promoting social inclusion. Apart from funding, the FEAD embodied the EU's commitment to reducing poverty and fostering social cohesion. The FEAD served as a crucial supranational tool for social protection, addressing urgent needs across EU countries by providing food and essential material assistance. It complemented national social protection systems, filling public food assistance gaps, and reinforcing the EU's role in combating poverty and deprivation. Through national programs, the FEAD relied on partner organizations, including public agencies, NGOs, and national food banks, such as members of the European Federation of Food Banks (FEBA), to deliver aid effectively to those in need. Founded in France in 1986 and now based in Brussels, FEBA (European Federation of Food Banks) is a key advocate in the European food aid policy. It represents 341 food banks in 30 countries; it aligns its efforts with EU programs such as the Multiannual Financial Framework, the European Pillar of Social Rights, the FEAD, and the European Social Fund Plus (ESF+). FEBA leverages its status as an EU lobby group to secure funding for essential infrastructure improvements in local food collection and distribution. Its advocacy on local infrastructural investments is vital, as certain studies highlight that food parcels from national food banks often fall short of nutritional standards, containing high-fat, sugary, and salty items that are inconsistent with dietary guidelines (Oldroyd et al., 2022). The main barriers to providing healthier meals include limited food storage, preservation funding, and the limited selection of food items received through donations. The FEAD addresses these issues by enabling food banks to deliver healthier and more sustainable food aid. European investments aim to improve the well-being of vulnerable populations while aligning with the EU's goals of social cohesion and resilience (Bazerghi et al., 2016; Oldroyd et al., 2022).

The main goals of the FEAD include alleviating extreme material deprivation by providing essential aid such as food, clothing, and basic goods to those in need, while also promoting social inclusion for marginalized individuals by integrating them into broader support measures and reducing food loss by redistributing surplus food. While the FEAD advances social citizenship and rights from the grassroots level, its reliance on NGOs, civil society organizations, and charities for food assistance highlights gaps in formal guidelines for the implementation and monitoring of programs (Oldroyd et al., 2022). The purpose of the FEAD was not to replace existing national public policies aimed at reducing poverty and social exclusion but to complement them by boosting and integrating assistance networks and interventions. Implementing the FEAD has varied across Europe and has been led by two distinct operational approaches. The first, adopted by Italy, France, Poland, and the United Kingdom, focuses on food and material aid, relying heavily on NGOs and charitable organizations, such as Food Banks, the Red Cross, Caritas, and OCMW Antwerp, to deliver essential assistance. The second operational approach, adopted by the Netherlands, Denmark, Germany, and Sweden, focuses more on a broader idea of social inclusion rather than on food or material aid. From 2016 to 2021, FEAD coordination followed two main pathways: one involving Member States' managing authorities, and the other involving street-level bureaucrats (SLBs) and NGO project operators, who actively participated in FEAD implementation at the ground level and played a pivotal role in enhancing capacities to tackle food and social vulnerability issues. The FEAD program was more than just a financial tool to channel EU funds to Member States. It was a platform for community building and collaborative exchange, it facilitated the acquisition and sharing of critical knowledge on implementation processes. The FEAD has enhanced networking, by building alliances between food aid organizations and public-private partnerships within national and local welfare systems. It has also promoted a rights-based approach to social support by facilitating *frequent interaction* among public-private street-level operators. The European central monitoring of best practices enabled street-level operators to exchange experiences, and gather feedback from other countries. It facilitated sharing insights on how public and private social workers and volunteers supported beneficiaries and improved their conditions on the ground. Some of the key challenges addressed were the need for more trained volunteers, better coordination of public NGOs, and removing other barriers to effectively reaching end beneficiaries. The FEAD increased cooperation between public and private actors across Europe, promoting mutual support, and the dissemination of best practices to enhance program outcomes. The European network's members kept in touch through an online community, social networks, and an ad hoc online platform (Yammer) where individuals could discuss key issues and share information and expertise. Between 2015 and 2020, the EC organized 18 network meetings and delivered thematic dossiers and catalogs collecting several national-urban case studies. The initiative of collecting and sharing FEAD best practices at the European level reflects a shift beyond the FEAD as a purely financial tool to address food and material support, toward a tool aiming to increase local capacity building for social policies. A key theme was how to design lasting, sustainable social inclusion strategies, focusing on how food and material assistance can complement broader social inclusion measures. This synergy seemed particularly relevant when it came to integrating food support with other social welfare initiatives, such as those funded through the European Social Fund (ESF). The FEAD Network started emphasizing the role of *'temporary food support as a critical entry point'* for designing more sustainable interventions that facilitate the beneficiaries' transition out of poverty. In urban contexts, during crises where food insecurity is often more acute owing to economic inequality and high costs of living, temporary food assistance can serve as a gateway for individuals to keep in touch with public-private collaborative networks and access social services. Thanks to this perspective, food aid is no longer considered a mere stopgap but rather a strategic element in broader social inclusion efforts (EU, 2017, p. 3). Urban food policy networks were therefore required to evolve, ensuring that food aid is integrated into a broader social safety net that promotes pathways to economic stability, employment, and social reintegration. It includes coordinating activities with municipal food programs, ESF-funded initiatives, and local social services to offer comprehensive, long-term support to the most disadvantaged urban populations.

From a multilevel perspective, the European Commission and the Member States also share governance accountability for FEAD programs. At the Member State level, Managing Authorities are tasked with certifying organizations to be granted the FEAD and conducting project evaluations while the Commission oversees the appropriateness of such assessments. Both levels rely heavily on data collected by national authorities. To support the Managing Authorities in their evaluation and monitoring responsibilities, the Commission has issued detailed guidance providing a framework and best practices to ensure that evaluations are thorough, consistent, and effective across different Member States. By following this guidance, managing authorities can make the necessary adjustments to improve their effectiveness in addressing poverty and social exclusion. Annual FEAD management and outcome reports are available to the public on the FEAD website. The monitoring of distribution and management of the Fund among the Member States for 2014–2020 takes into equal measure the AROPE index. With the adoption, in June 2023 of Regulation (EU) No. 2021/1057 European Social Fund Plus (ESF+), the ESF community of practices of material support was activated, and the European Commission approved the Member States' National programs to combat food poverty.

2.4 Closing Local Gaps: Street-Level Implementation of FEAD Programs

In response to the urgent challenges exacerbated by the COVID-19 pandemic, in April 2020, the European Commission amended the Fund for European Aid to the Most Deprived (FEAD) as part of the Coronavirus Response Investment Initiatives Plus (CRII+) package. This initiative aimed to enhance flexibility and liquidity for Member States. Further adjustments were proposed in May 2020 to the Common Provisions Regulation and FEAD Regulation, allocating additional resources for 2020 through 2022 and bolstering the cohesion policy within the EU's 2014–2020 budget framework. By February 2021, amendments to the FEAD Regulation were implemented to enable financial transfers from REACT-EU, reinforcing the Fund in response to the ongoing COVID-19 crisis. The compounded impacts of the pandemic and the Russia-Ukraine conflict again prompted European institutions to redefine the role of the FEAD together with other European funds, emphasizing greater flexibility. In April 2022, legislative amendments adopted by the Council empowered Member States to redirect resources from cohesion policy funds and the FEAD to support refugees from Ukraine. This move underscored the EU's solidarity with affected Member States, mainly those sharing borders with Ukraine.

Additionally, the FEAD's mandate expanded to provide essential assistance, including food and basic materials, to individuals affected by Russia's military actions, encompassing third-country nationals. The CARE legislation facilitated swift adjustments to operational programs supported by the FEAD within Member States, bypassing the usual requirement for Commission decisions. Specific programs devoted to segments of vulnerable populations, such as Child Guarantee, adopted by the EU in 2021, were adopted to ensure the protection of the rights of all children and secure access to essential services and healthy meals each school day.[3] Ensuring access to at least one healthy meal each school day (a broader concept than access to school meals) is one of the objectives of the European Child Guarantee program aimed to ensure that every child in Europe at risk of poverty or social exclusion has access to the following vital services: (1) early childhood education and care, (2) education (including school-based activities), (3) healthcare, (4) nutrition, (5) housing.

The COVID-19 initiatives introduced unprecedented flexibility by relaxing conditionality clauses, allowing Member States to reallocate unspent resources from the 2014–2020 programming period of the European Structural and Investment Funds (ESIF) and the Fund for European Aid to the Most Deprived (FEAD),[4] significantly enhanced the FEAD's role in mitigating the impacts of the COVID-19 crisis.

It enabled the rapid implementation of anti-crisis measures to address the urgent needs of the population. Compared to the 2008–2009 financial crisis, the EU's response to COVID-19 was significantly faster and broader in scope. The emphasis on monitoring and reporting ensured transparency and traceability of operations and outcomes, while ad-hoc working groups and interinstitutional mechanisms streamlined decision-making and minimized the duplication of efforts. Furthermore, the pandemic highlighted the need for enhancing collaboration with social partners to ensure that interventions aligned more closely with on-the-ground needs. This experience underscored the importance of flexibility,

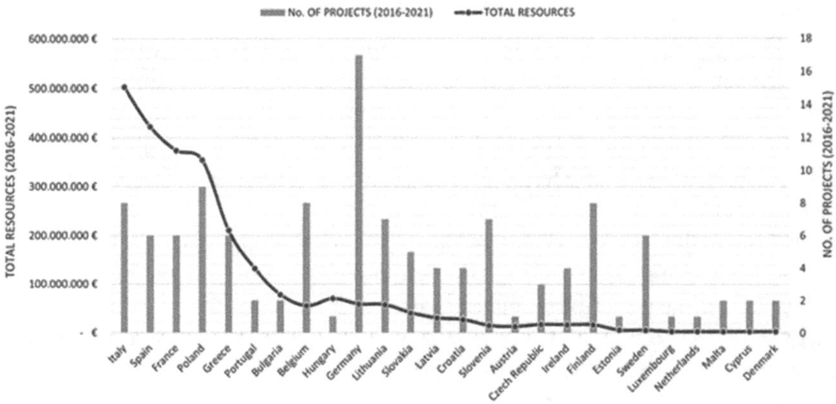

Fig. 2.2 Countries' total available FEAD resources for the years 2016–2021 (mm) and the total number of projects carried out in the same years. (Source: Cohesion Open Data Platform, 2021, https://cohesiondata.ec.europa.eu/stories/s/In-profile-Fund-for-European-Aid-to-the-Most-Depri/tdry-xg55/ and European Union FEAD case studies 2016, 2017, 2018, 2019, 2021)

coordination, and inclusivity in addressing socio-economic crises efficiently and effectively (EC-DG Employment, Social Affairs and Inclusion, 2023).

Despite additional funding and diverse contexts and conditions, the FEAD guaranteed each Member State a minimum allocation of EUR 3500,000 for the 2014–2020 programming period, ensuring a significant resource base for developing operational programs.

As shown in Fig. 2.2 (blue line), the overall allocation of FEAD funds across European countries during the 2014–2020 period (comprising 2021 for financed project conclusion) predominantly favors some Member States, notably Italy, France, Spain, Romania, Poland, and Greece (Fondazione Brodolini, 2021). Before the outbreak of COVID-19, the three major contributors to the overall number of end recipients of food assistance in the EU in 2019 were, unsurprisingly due to their significant financial allocation, France (39.3%), Italy (17.1%), and Poland (11.0%), closely followed by Spain (10.1%). Poland entered the EU in 2004. Despite not being among the countries with the highest vulnerability rate, Germany had the most significant number of best practices reported by the EC, followed by Poland, Italy, Belgium, and Finland.

Italy, France, Poland, and Spain received the largest share of funds and have the same number of best practices analyzed related to the food-aid-oriented FEAD OP1. In contrast, while having fewer financial resources, Germany shows a higher number of case studies mainly focused on social integration, where food is a minority component (FEAD OP2). The table below illustrates the AROPE trends from 2014 to 2020 for the five European countries that receive the most financial support from the FEAD. All Member States, except France, experienced a reduction in the AROPE index in the period under examination. While FEAD funds are thought to have contributed to reducing AROPE levels in Europe, the causal relationship between financial aid and the vulnerability reduction is not straightforward, and it is challenging to separate the specific impact of FEAD monetary transfer from that of other European or national funds or programs aimed at alleviating food poverty and insecurity at the local level (Fig. 2.3).

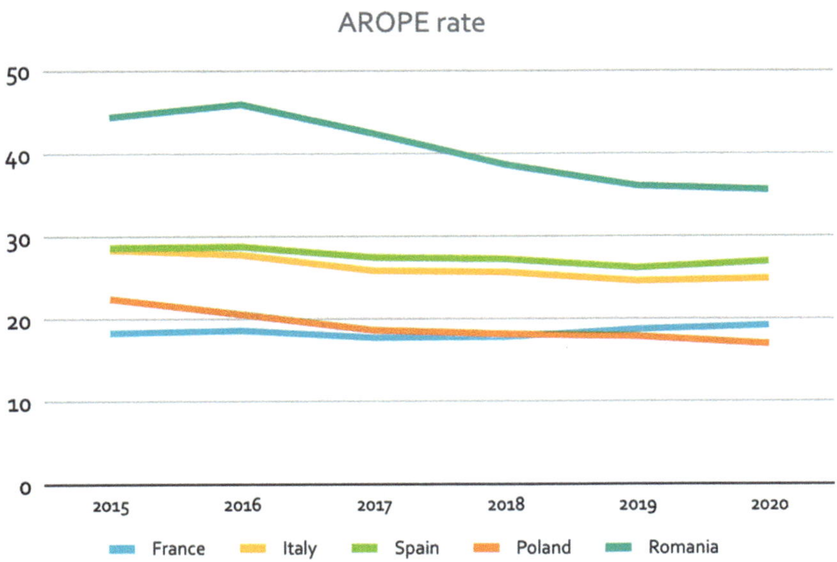

Fig. 2.3 The AROPE rate trends (2014–2020) for the five European countries receiving the most financial support from the FEAD. (Source: Made by the author based on Eurostat (2024), https://ec.europa.eu/eurostat/statistics-explained/index.php?title=Living_conditions_in_Europe_-_poverty_and_social_exclusion#Source_data_for_tables_and_graphs)

Based on the estimations by partner organizations, the FEAD supported nearly 13 million persons per year, between 2024 and 2020, receiving food support and/or material assistance; most of them were children. To investigate deeper local governance dynamics of food assistance, the explorative multi-year analysis of FEAD projects illustrated in the following sections analyzes the execution of food programs at the street level, the type of SLBs involved, and their collaborative activities. Our database covers 125 out of 133 FEAD best practices published in the catalogs released in the framework of the mutual learning activities of the FEAD Community in 2016, 2017, 2018, 2019, and 2021 (FEAD activities were suspended in 2020, and food aid was delivered under the CRII and CRII+). The case studies were identified by the Mutual Learning Services team supporting the FEAD Community and selected by the European Commission based on the following criteria: relevance, involvement of different actors, effectiveness, and transferability (FEAD, 2021). The catalogs showcase initiatives helping individuals transition toward social inclusion (FEAD, 2021). The database includes key variables such as project titles, publication years, and chosen Operational Programs (OPs): 'Type I,' focused on food aid and material support and 'Type II,' addressed social inclusion policies for the most deprived, supplemented by qualitative data from project descriptions. The governance features included in the process of delivering services to vulnerable people are grouped into three macro-dimensions: *The rationale of food security governance, the actor configurations and multilevel diversity, and the beneficiaries' interactions.*

The Rationale of Food Security Governance

The first macro group of variables investigates how best practices framed food assistance in poly-crises. First, how food aid in FEAD projects is designed and how implementation strategies intersect with other policy fields. Each project identifies a primary policy field that reflects the central focus of the intervention and other various policy fields including food poverty, education, empowerment, governance, health, homelessness, housing, immigration, psychological support, and training. The more intersectionality there is, the more one can expect the multidimensionality of networking capacity and integration to increase. Second, by classifying projects as 'principal,' 'supplementary,' or 'unrelated' to food security, one can determine the extent to which best practices prioritized food aid as the entry point for addressing the social integration of the vulnerable. Thirdly, projects to highlight whether interventions are directly linked to the pandemic or other crises to set up measures and mitigate food insecurity and social exclusion. Only explicit references to crises were considered.

The content analysis examines the intersectionality of FEAD best practices, how food aid is positioned and integrated within local design and implementation strategies, and what projects adopt a more holistic approach to addressing the diverse needs of the most deprived populations. The content analysis of all the 125 project reports considered, conducted using the IRAMUTEQ program, identified several dominant clusters of topics:

(a) Food distribution and delivery through charity organizations, food banks, and civil society organizations (dark blue cluster);
(b) Service counseling for accommodation and housing (grey cluster);
(c) Food distribution in schools, donations (light blue cluster);
(d) Inclusion and empowerment of people through participation and emotional promotion (red cluster);
(e) Vulnerability and employment assistance, disability, older adults (green cluster);
(f) Workshops and learning for volunteers and organizers (purple cluster) (Fig. 2.4).

Since the projects selected by the EC are considered best practices in assisting vulnerable individuals and fostering local-level learning, it is unsurprising that all the emerging clusters of themes—corresponding to activities delivered—are interconnected. Notably, the 'food distribution and delivery through charity organizations, food banks, and civil society organizations' (dark blue) and 'food distribution in schools, donations' (light blue) clusters show particularly strong linkages. This reflects the fact that a significant portion of those assisted by FEAD projects are children, who rely heavily on school-based food programs and charitable distributions to meet their nutritional needs. They are also closely related to the 'workshops and learning for volunteers and organizers' (purple) and 'inclusion and empowerment of people through participation and emotional promotion' (red) clusters. What is new is the role of food assistance, both public and through intermediaries, as a key driver of integration and community resilience, helping local communities better address social vulnerability. In some countries, food assistance is more closely tied to specific issues. For example, in Germany, it is often part of projects supporting

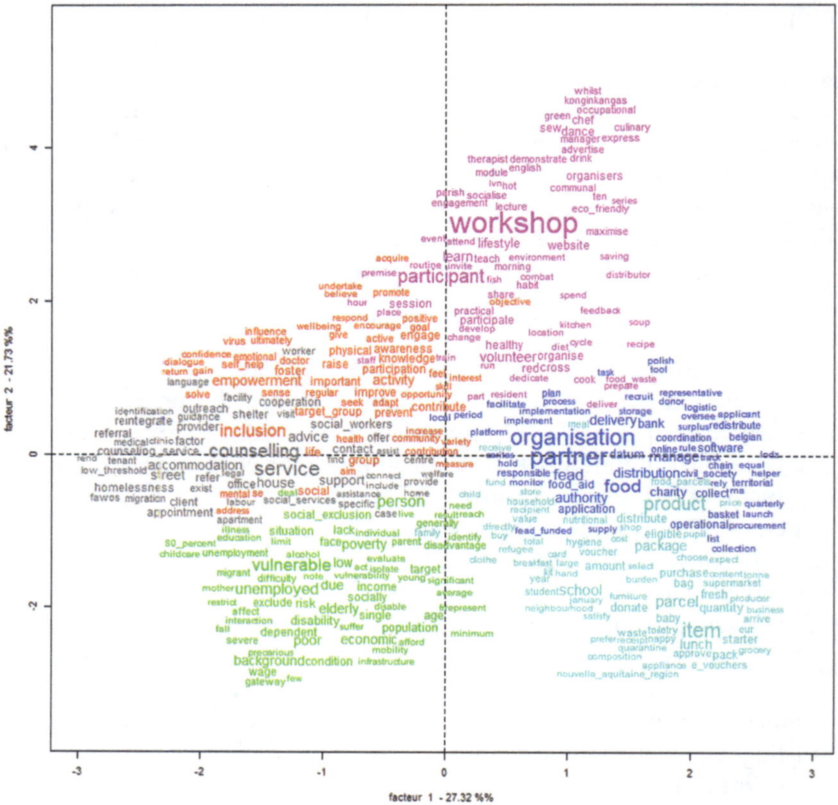

Fig. 2.4 Interlinks between topics in FEAD projects. (Source: Made by the author based on European Union FEAD case studies (2016, 2017, 2018, 2019, 2021a, 2021b))

vulnerable groups such as immigrants and people experiencing homelessness. In contrast, countries such as Czechia, Hungary, Austria, Belgium, and Poland focus more on combating food poverty among schoolchildren. These projects aim to ensure that children from low-income households have access to adequate nutrition, emphasizing the role of schools as key sites for food assistance. Poland also has initiatives where food banks and volunteers collect and redistribute surplus food to vulnerable populations. These organizations play a critical role in collecting surplus food and redistributing it to various vulnerable groups, including families, the

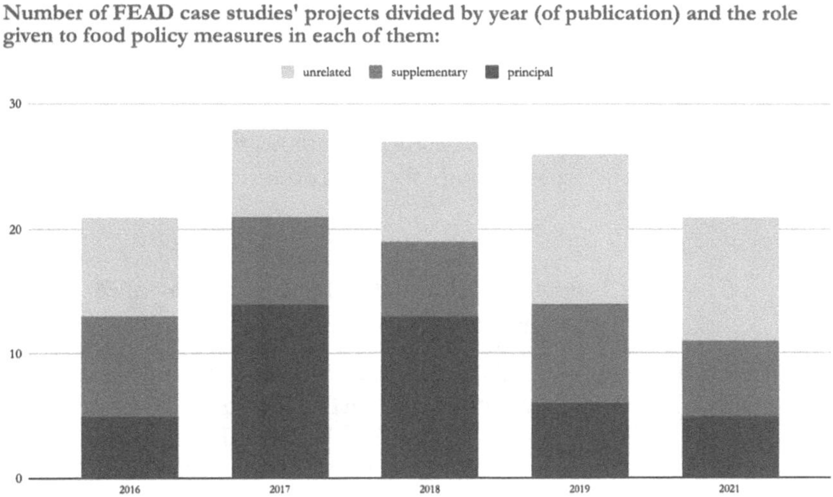

Fig. 2.5 FEAD projects based on the relevance of food security measures (2016–2021). (Source: Made by the author based on European Union FEAD case studies (2016, 2017, 2018, 2019, 2021a, 2021b))

elderly, and people experiencing homelessness. Driven by civic engagement and the efficient use of food resources, this model of food assistance has become a key part of Poland's strategy to address food insecurity. In the absence of a specific national food security policy, Poland shows the potential of community-based solutions to tackle widespread vulnerability through local public and volunteer-driven initiatives.

Figure 2.5 illustrates that food aid measures are consistently paired with other forms of intervention to establish the integrated nature of vulnerability and local intervention. In projects carried out in 2017–2018, food aid was the dominant measure, whereas, during the pandemic, there was a more balanced distribution of food aid and additional support measures. Only 3 projects out of 125 mention the financial crises, and 11 explicitly mentioned the COVID-19 pandemic. For example, during the COVID-19 lockdown in March 2020, the Centre for Regional Universities and Schools (CROUS) in the Nouvelle-Aquitaine region (France) identified students in third-level education who could benefit from food parcels. Food banks delivered a weekly 8 kg package to each identified student, including dry and fresh food products and hygiene kits (EU, 2021). In

Germany, the Social Bridge Hamburg project supports mobile citizens from EU Member States (primarily Bulgaria, Poland, and Romania) living in precarious conditions in Hamburg, especially during the COVID-19 pandemic. The project established a local network of organizations providing practical support and counseling, and connecting beneficiaries with other suitable support services (e.g., food distribution points and clothing stores) (EU, 2021).

2.5 Actors on the Frontlines: Who Shapes FEAD Programs

In mapping the range of public actors and levels of government involved in the street-level implementation of FEAD projects, a first distinction concerns the number and types of Street-Level Bureaucrats (SLBs) engaged. These include not only the traditional categories identified by Lipsky (1980), such as social workers, police officers, and teachers, but also an expanded group of professionals. Municipal government officials make up the most numerous group involved in the project, often working alongside social workers, healthcare professionals, and representatives from educational institutions, such as universities and schools. Social workers play a crucial role in fostering inter-organizational learning within the city project. Their work ensures that knowledge, best practices, and insights are continuously shared across institutions, strengthening collaboration and improving service delivery. The FEAD projects undergo annual evaluations, with findings used to enhance services. In decentralized administrative systems, a key measure of success is the high rate at which users follow through on referrals to existing welfare institutions, as seen in Berlin (FEAD, 2021). A significant proportion of users connect with social workers from counseling services, where they receive status verification, enabling access to long-term case management. However, in cases where status verification is centrally administered by social operators—such as in Poland's Social Welfare Centers—street-level operators may face challenges in inter-organizational networking and learning. To address this, the FEAD Managing Authority introduced a new procedural option for food assistance registration. The Ministry for Family, Labor, and Social Policy implemented a decentralized mechanism allowing beneficiaries to initiate the verification process through partner organizations providing food assistance. These organizations then forward the necessary forms to Social Welfare Centers, which remain responsible for verifying applicants' eligibility. This procedural innovation not only streamlines the verification process but also helps overcome the stigma that often discourages people in need from registering at state facilities (FEAD, 2016). Originally introduced in Poland, to address the negative effects of the economic crisis and the Russian invasion of Crimea, this reform significantly improved food assistance efficiency and capacity, proving especially valuable during the pandemic crisis. In Bulgaria, Partner Organizations played a crucial role in distributing FEAD-funded food parcels to end recipients. However, their impact extended beyond mere food distribution. Volunteers and social workers used informal processes as an opportunity to engage with beneficiaries informally, fostering trust and open communication. The informal interaction enabled them to identify individuals who might need additional support, whether related to housing, employment, healthcare, or social services (FEAD, 2017). During the pandemic, informal and digital procedures and initiatives are particularly effective in reaching vulnerable populations who were hesitant to seek assistance through formal channels (FEAD, 2021). In Greece, through the 'We Strong Together' project, both the telephone support line and the online platform played a crucial role in reducing the social stigma associated with COVID-19. This was largely due to the inclusive nature of these services, which were accessible to everyone, not just vulnerable groups. Additionally, operators tested a more neutral and less stigmatizing form of communication and terminology, further encouraging individuals to seek support without fear of judgment (FEAD, 2021).

The great variety of organizations involved underscores the diverse range of public service sectors and actors involved in the implementation and administration of FEAD projects. While this variety aligns with the types of SLBs commonly discussed in the literature, there is a notable exception in the case of 'city government officials'. This group shows more extensive engagement with stakeholders, volunteers, and policy recipients in addressing social, economic, and food vulnerabilities than expected.

Some interesting patterns emerge when examining the distribution of SLBs participating in the local implementation of FEAD projects. Germany, with its highly decentralized policymaking system, is particularly noteworthy for having the highest number of SLBs involved and for the diversity of these roles. Countries such as France, Bulgaria, Greece, Croatia, Latvia, and Lithuania also exhibit a significant variety of participating SLBs, thus indicating a multifaceted approach to managing FEAD projects.

In Finland and Poland, projects involved a much narrower range of local operators. This more limited engagement suggests a more centralized social system or streamlined approach to project management reflecting different administrative priorities or resource allocations. Overall, the data highlights that best practices are defined by the diverse and innovative strategies and procedural tools employed by different countries in mobilizing SLBs to tackle food insecurity and related social challenges.

Despite the variation in roles, most SLBs work in the local implementation of food aid programs, frequently participating in the same projects to ensure effective coordination and delivery of services. In project execution, SLBs contribute to the following activities: (1) link to the project (e.g., when SLBs determine eligibility or refer the target group to a project), (2) link from the project (when the project helps the target group access services provided by SLBs, thereby referring the group to SLBs), (3) project administration (when SLBs are formally involved in project management), and (4) project implementation (when SLBs cooperate with or support the project's daily activities). The SLBs are distributed along different measures (Fig. 2.6). Most intervene in projects devoted to food and material assistance, social inclusion, and counseling measures (Fig. 2.7).

The substantial scope of intervention by country is illustrated in the figure below ('Number of FEAD case studies' projects by country and type of social policy field'). The figure highlights that some case studies in Croatia, Cyprus, Luxembourg, and Portugal focused exclusively on food aid programs. By contrast, other case studies integrated food assistance with a broader range of social support services and sectors. Italy, Spain, and Germany demonstrate the greatest diversity in the sectors and tools they use, followed closely by Belgium, Finland, and Poland. This variety underscores the need for a comprehensive approach to addressing vulnerability—one that integrates efforts to combat food poverty with broader social assistance initiatives. By adopting multifaceted strategies, these countries enhance support systems for vulnerable populations, ensuring a more holistic and effective response to social exclusion (Fig. 2.8).

When examining the activities of street-level bureaucrats (SLBs), it becomes clear that a significant majority are involved in implementation tasks. In 39% of cases, SLBs primarily focus on the practical execution of policies and programs. This includes managing application

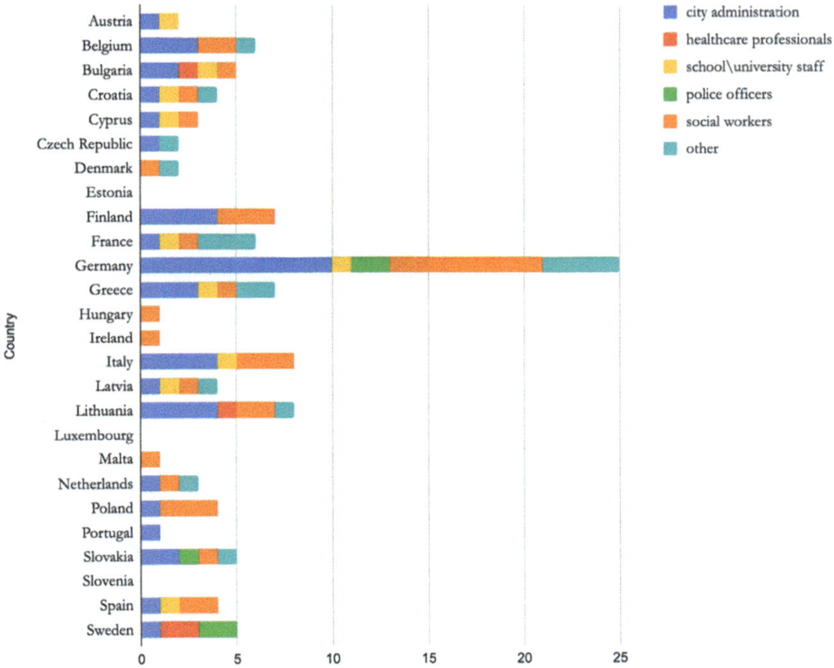

Fig. 2.6 Number of street-level bureaucrats identified in FEAD case studies' projects divided by country and by typology. (Source: Made by the author based on European Union FEAD case studies (2016, 2017, 2018, 2019, 2021a, 2021b))

procedures—both digitally and in person—for food aid distribution, as well as overseeing public food procurement to ensure the quality and sustainability of food products (FEAD, 2016, 2017). Their work is hands-on and central to the successful delivery of public services and aid, highlighting the importance of SLBs in translating policy into tangible outcomes.

Projects can be categorized based on the scale of their interinstitutional collaboration, ranging from more centralized to decentralized structures. **Local projects** typically have an urban or suburban scope, involving one or more cities or local authorities. These initiatives tend to focus on addressing specific community needs through localized governance and close cooperation between municipal agencies, social services, and civil society organizations. **Multilevel projects,** on the other hand, involve actors from multiple geographical levels, such as state, regional, and

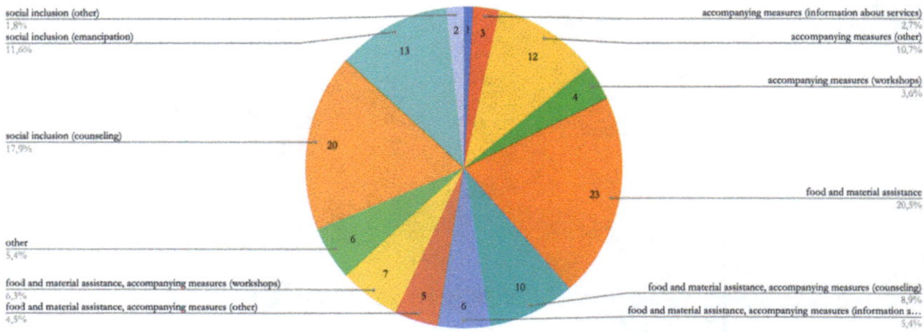

Fig. 2.7 Total number of street-level bureaucrats identified in FEAD case studies' projects divided by the type of measures. (Source: Made by the author based on European Union FEAD case studies (2016, 2017, 2018, 2019, 2021a, 2021b))

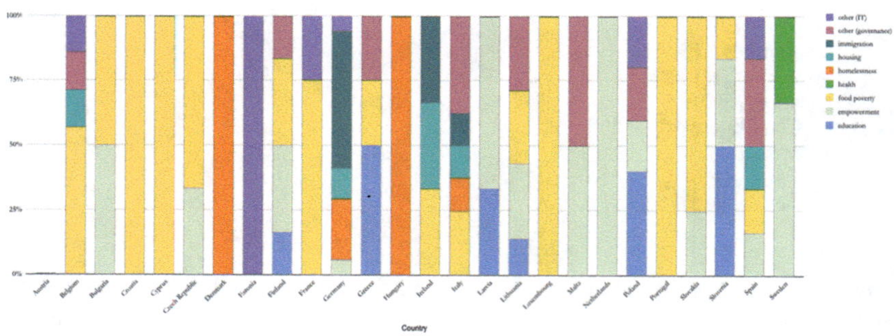

Fig. 2.8 Number of FEAD case studies' projects divided by country and the type of social policy field. (Source: Made by the author based on European Union FEAD case studies (2016, 2017, 2018, 2019, 2021a, 2021b))

municipal governments working together. These projects require a more structured and often centralized coordination framework to ensure alignment between different administrative levels. The complexity of managing interinstitutional collaboration at multiple levels can pose challenges, requiring robust governance mechanisms to maintain efficiency and responsiveness.

The implementation of FEAD projects varies across European countries, reflecting different levels of administrative centralization and governance structures. In some countries, such as Germany and Italy (Fig. 2.9), implementation activities are predominantly carried out by local-level

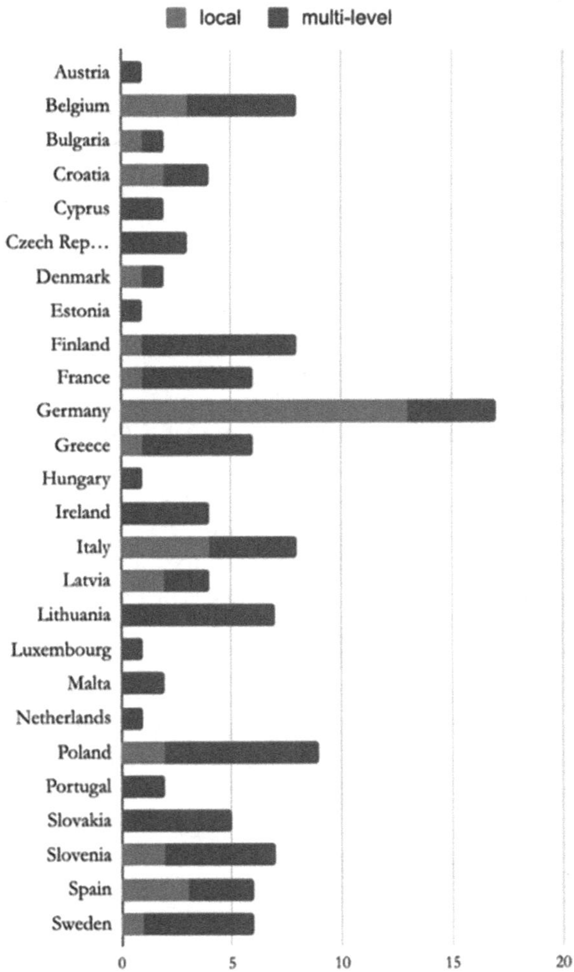

Fig. 2.9 FEAD case studies' projects divided by country and by the governance level that plans and oversees each project. (Source: Made by the author based on European Union FEAD case studies (2016, 2017, 2018, 2019, 2021a, 2021b))

administrations and entities. These decentralized approaches enable municipalities and regional authorities to tailor interventions to local needs, ensuring flexibility and proximity to beneficiaries. In contrast, in countries such as Poland, Belgium, Bulgaria, Denmark, Finland, France, Slovakia, Spain, Greece, Sweden, and Latvia, the central administration plays a significant role in managing FEAD projects. Here, national authorities coordinate funding distribution and program oversight while working with regional and local actors to implement services. Meanwhile, in several other countries, the implementation of FEAD projects is exclusively overseen by central government units operating at the territorial level. This is the case in the Netherlands, Estonia, Croatia, Hungary, Lithuania, Malta, Ireland, Austria, Cyprus, Czechia, Luxembourg, Portugal, and Slovakia. These highly centralized systems ensure standardized procedures and uniform service delivery across regions but may have less flexibility in adapting to local contexts. This variation in administrative responsibility highlights the diverse governance models, jurisdictional structures, and strategic approaches adopted by European countries in managing and carrying out FEAD projects.

Implementing FEAD best practices is rarely managed by specialized units focused solely on food assistance. Instead, food aid is typically integrated into a holistic governance approach to addressing poverty, social and food vulnerability, where food assistance is just one element of a larger strategy to improve overall social welfare. Integrating different types of assistance highlights the multi-functional role of SLBs, who are not only implementers of specific aid programs but also key coordinators of a range of services that contribute to social inclusion. Their work requires navigating complex social issues, collaborating across sectors, and adapting to diverse needs and contexts. Effective cross-sector collaboration among SLBs is critical for ensuring that food assistance is complemented by other forms of support, such as housing, healthcare, employment services, and educational opportunities. Having recognized the need for comprehensive support, the European Commission has encouraged the networking of SLBs from various sectors, fostering local collaboration to create more cohesive and supportive environments for those in need. The interactions between SLBs and municipal authorities during the implementation phase of FEAD projects emphasize the crucial role of local government officials in delivering practical social assistance. Whether local or state-level, SLBs primarily engage with municipal authorities, underscoring the importance of localized, flexible, and integrated approaches to service delivery. These dynamics highlight the need for capacity-building and strong intergovernmental collaboration to ensure the success of food aid and related programs. Recognizing and addressing these factors during crises can significantly enhance the impact, sustainability, and long-term benefits of FEAD and similar initiatives to alleviate poverty and promote social inclusion.

2.6 Beneficiaries' Voices: Navigating the Food Aid System

The beneficiaries of FEAD projects, intermediaries (such as NGOs and local agencies) and final recipients of food aid and material assistance play a major role in shaping street-level governance dynamics. Article 13(6) of Regulation (EU) No. 223/2014 outlines specific categories of eligible end beneficiaries (target groups) for FEAD projects, applicable across all European Member States: (1) homeless individuals; (2) migrants, participants with a foreign background, and minorities; (3) women; (4) children aged 15 or younger; (5) persons aged 65 or older; and (6) persons with disabilities. Figure 2.10 shows that the general target groups, 'food aid recipients' and 'people in economically precarious conditions' are the most common beneficiaries indicated by the projects, followed by migrants, other categories (women, students, elderly people), and children aged 15 or younger.

The Composition of the Target Groups by country (Fig. 2.11) illustrates the groups prioritized by food aid programs, categorized by

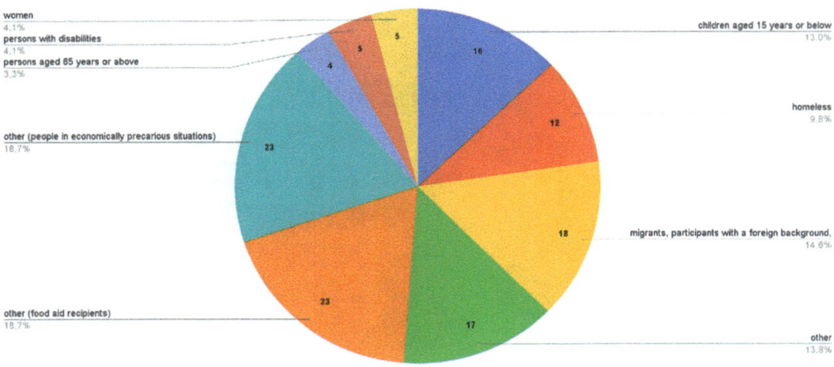

Fig. 2.10 Target group categories of FEAD projects. (Source: Made by the author based on European Union FEAD case studies (2016, 2017, 2018, 2019, 2021a, 2021b))

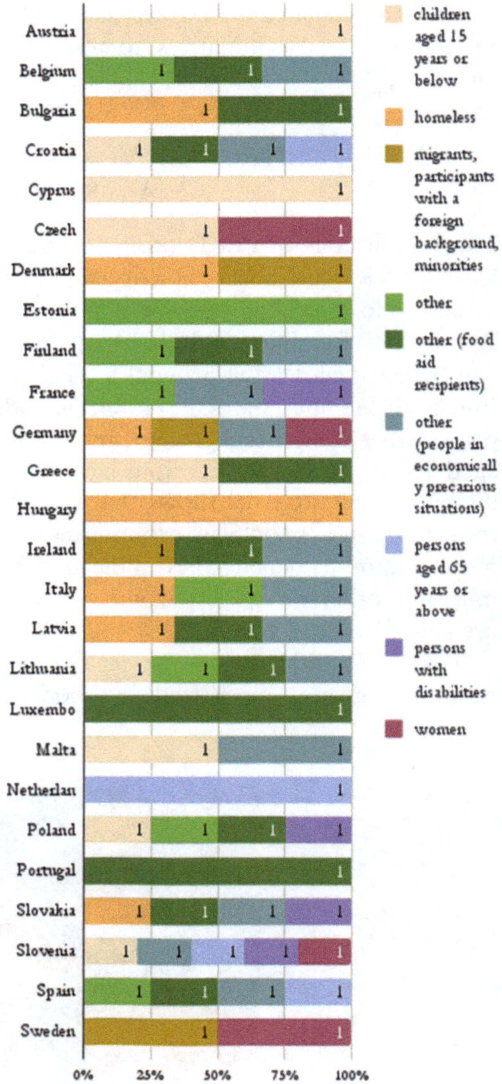

Fig. 2.11 Composition of target groups by country. (Source: Made by the author based on European Union FEAD case studies (2016, 2017, 2018, 2019, 2021a, 2021b))

country, and addressed at the street level. This figure illustrates the diverse population segments reached by food aid and material assistance across different national contexts, highlighting variations in target group definitions and the beneficiaries of FEAD projects. It also highlights the way public and private street-level operators shape the concept of vulnerability, reflecting the unique social, economic, and cultural factors that influence food aid distribution in each country.

The evidence indicates that certain countries narrowly focus on specific, homogeneous target groups when addressing food insecurity and vulnerability. For example, Austria and Cyprus prioritize children in their food aid initiatives, while the Netherlands concentrates on individuals aged 65 and older. By contrast, Hungary directs its efforts toward supporting the homeless population. On the other hand, countries such as Slovenia, Spain, Slovakia, Croatia, Lithuania, Germany, and Poland exhibit a broader diversity in their target groups. This diversity shows a more comprehensive approach to combating food insecurity, as it simultaneously recognizes and addresses the needs of various segments of society. Variations in the composition of target groups suggest that street-level operators in different countries exercise discretion in the recognition process (Hermans & Cantillon, 2025). This discretion influences the inclusiveness of food aid programs, enabling operators to tailor interventions to local needs. By recognizing the complex relationship between food insecurity and other forms of social disadvantage, they can adopt a more integrated approach that combines food aid with other forms of assistance. Serving diverse target groups fosters stronger collaboration within implementation communities, enabling multiple areas of intervention and a wider range of support tools. This integrated approach facilitates knowledge-sharing, resource pooling, and improved outcomes, ultimately leading to more effective and responsive interventions.

The analysis underscores the importance of understanding target group composition in food aid programs. More inclusive implementation processes provide comprehensive support, thus enhancing social cohesion and community resilience. Additionally, the role of street-level operators is crucial, as their decisions directly impact the accessibility and effectiveness of food aid efforts across Europe. By prioritizing diverse target groups, these operators can better address the complexities of food insecurity, tailoring aid programs to the unique challenges faced by each group, and increasing the overall impact of food aid programs. Before the pandemic, and until 2019, women made up just under half (49.5%) of all individuals assisted by the FEAD. Children accounted for 30%, while migrants, people with a foreign background, or minorities accounted for 10%. Additionally, 7% of end recipients were homeless, 8% were aged 65 or older, and slightly less than 5% had disabilities. These categories are not mutually exclusive (Fondazione Brodolini, 2021, p. 16). Given the differences in accompanying measures across countries and projects, the type of assistance has been classified based on the European Court of Auditors' special report on the FEAD. The categories include:

- **Food and material assistance**—such as food packages, cooked meals provided in social canteens, or distribution through partner organizations.
- **Accompanying measures**—including information about services, workshops, and counseling.
- **Social inclusion initiatives**—aimed at fostering long-term support and integration.

Distinguishing these types of assistance provides a clearer understanding of how food security measures are implemented and their role in promoting social inclusion and stability (Fondazione Brodolini, 2021, pp. 18–19).

An annual analysis of target groups (Fig. 2.12) reveals a clear distinction between the periods before and during the COVID-19 pandemic in

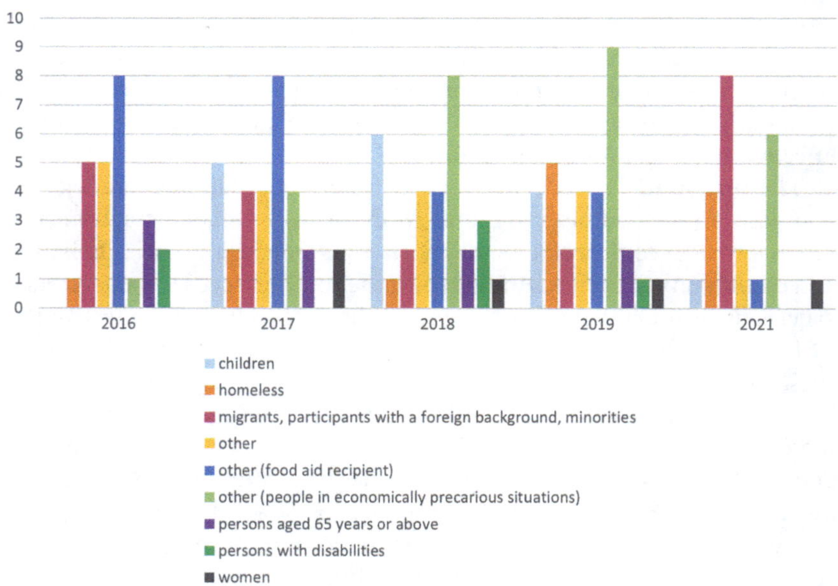

Fig. 2.12 Target groups by year. (Source: Made by the author based on European Union FEAD case studies (2016, 2017, 2018, 2019, 2021a, 2021b))

2020. Notably, the number of migrants and individuals with a foreign background receiving aid increased significantly during the pandemic. This shift was largely due to the economic crisis and widespread job losses, which disproportionately affected precarious workers and the most vulnerable segments of the labor market. The data also highlights that homeless individuals, women, and migrants were the primary beneficiaries of projects focused on service counseling. This reflects the heightened need for social support and guidance during the pandemic, helping individuals navigate available aid measures. Meanwhile, children remained the main focus of FEAD-funded projects related to food management and distribution, underscoring ongoing efforts to combat child hunger and malnutrition. Beyond shifting aid priorities toward vulnerable populations such as migrants and minorities, the pandemic reinforced the need for targeted interventions. Counseling services became essential for those facing severe social and economic hardships, while food distribution efforts remained crucial for children. These changes illustrate the adaptability of social assistance programs in response to crises, emphasizing the importance of flexible and responsive strategies to address poverty and deprivation. After the pandemic, the proportion of aid for children declined, largely due to the introduction of specific ESF+ provisions for minors, such as the Child Guarantee.

Key stakeholders involved in FEAD projects include actors from the private sector, public agencies, third-sector organizations (both for-profit and non-profit organizations), religious bodies, and other contributors, such as independent volunteers (e.g., lawyers and students). These stakeholders offer diverse resources, expertise, and perspectives, enabling more coordinated and effective responses to food insecurity. However, the pandemic posed formidable challenges for public and private actors, such as voluntary organizations like the Red Cross, Food Banks, and Caritas. In response to rising operational challenges during the pandemic, the introduction of electronic food vouchers, though potentially empowering individuals by giving them greater autonomy over their food choices, has transformed governance modes both for public operators and charitable and non-profit organizations. This shift required public and private organizations to adopt innovative strategies to deliver food aid in the face of growing demand. The following figure (Fig. 2.13) illustrates the distribution of stakeholders and street-level bureaucrats (SLBs) involved in FEAD projects by country, that is, a visual representation of their roles across different countries. This data offers valuable insights into the operational

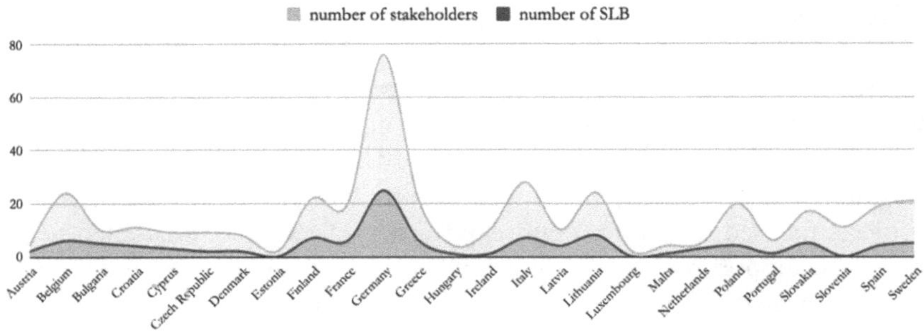

Fig. 2.13 Total number of stakeholders and street-level bureaucrats mentioned in FEAD case studies' projects divided by country. (Source: Made by the author based on European Union FEAD case studies (2016, 2017, 2018, 2019, 2021a, 2021b))

structure of FEAD projects in various countries. Figure 2.13 shows that Germany has the highest number of stakeholders and street-level bureaucrats (SLBs) involved in FEAD best practices, followed by Italy, Lithuania, and Finland. This proves that the third sector and public intervention are not mutually exclusive; rather, they can collaborate as equal partners to enhance social support efforts.

As anticipated by the European Commission, public-private collaboration and the involvement of diverse actors play a key role in shaping collaborative learning strategies. This, in turn, enhances the effectiveness of local food aid initiatives, aligning with EU expectations.

Germany's approach to FEAD project implementation is marked by extensive collaboration and broad engagement at the community level. The strong involvement of stakeholders and street-level bureaucrats (SLBs) reflects a well-coordinated national effort to combat food insecurity and social exclusion. Similarly, Italy, Lithuania, and Finland show a strong commitment to mobilizing resources and local expertise to effectively manage and deliver aid. This pattern suggests a more decentralized governance model for achieving the FEAD's objectives across these countries.

Germany's role in FEAD best practices is particularly notable. Despite relatively low funding, it leads in the number of case studies collected by the European Commission and in the number of cities (19) participating in FEAD projects. Before the pandemic, Germany's widespread system of food banks and civil society organizations played a crucial role in promoting food security, often collaborating with educational institutions and immigration offices. Food poverty was already a major issue, in particular, among vulnerable children and adolescents from lower socioeconomic backgrounds, where poor nutrition was more prevalent (WBAE, 2023). Likewise, local food councils in Germany have taken on a pioneering role in tackling food insecurity, following a bottom-up, participatory, and multi-stakeholder approach. This model engages civil society, food banks, farmers, and, to a lesser extent, public institutions. However, the COVID-19 crisis posed significant challenges. During the first wave, nearly half of Germany's 949 food banks (Tafeln) closed, with around 120 still closed by mid-2020. Their reliance on volunteers made them particularly vulnerable to disruptions caused by the pandemic (Simmet & Stroebele-Benschop, 2023). In response, new initiatives emerged in collaboration with municipalities to address food distribution challenges. Although the German federal government allocated over €150 billion to mitigate the pandemic's economic and social impacts, food security was not a specific focus of its relief measures. Instead, policies primarily addressed household income issues, and lacked the holistic approach to vulnerability encouraged by the European Commission (WBAE, 2023). Nevertheless, city-level initiatives during the pandemic underscored the importance of bottom-up, adaptive strategies, community involvement, and intersectoral collaboration. By 2021, approximately 1.65 million people in Germany relied on food banks, which redistributed consumable food to socially and financially disadvantaged individuals. These distribution centers, often managed by non-profit organizations in partnership with city governments, provided essential groceries and meals. Germany's FEAD best practices stand out for the strong engagement of food banks, civil society organizations, and local stakeholders grouped in food councils, reflecting intense activism at the local level. However, food insecurity remains a challenge. According to Eurostat data from 2022, 22% of Germany's lower-income population—particularly migrants—still cannot afford a nutritious meal every other day.

This work's analysis is consistent with findings from a recent comparative study on third-sector organizations across Belgium, Germany, Hungary, Lithuania, the Netherlands, Poland, Portugal, and Spain, which reveals the diverse connections between food aid and urban welfare policies (Greiss et al., 2022). The study highlights the diverse connections between food aid and urban welfare policies (Greiss et al., 2022) and points to a growing trend toward the institutionalization of food aid in

Europe, largely supported by the Fund for European Aid to the Most Deprived (FEAD) before and during the pandemic. By focusing on non-profit entities that directly offer food aid, the comparative study reveals several key trends: volunteer organizations often receive support from governmental actors and operate within a broader network of stakeholders; collaboration between non-governmental organizations (NGOs), charities, non-profits, and government bodies in providing food aid is becoming increasingly common; volunteer organizations frequently work with municipalities and local public authorities, playing active roles in food aid programs; many volunteer organizations act as intermediaries, requiring proof of need from individuals seeking assistance or referrals from public institutions (Greiss et al., 2022).

2.7 Mastering FEAD Governance: Proven Mechanisms for Effective Implementation

This work's analysis of aggregated best practices reveals that the European Union has prioritized strengthening key structural and functional aspects of food aid initiatives. A core best practice, in terms of FEAD implementation, is the adoption of an *integrated* approach to combating food insecurity as a critical component of addressing social vulnerability. Rather than relying on a rigid governance model, this approach remains *flexible*, *adapting* to institutional contexts while *fostering dialogue* among key stakeholders. *Collaboration* between public institutions, private sector actors, and empowered civil society is central to this model, as it ensures that support systems build upon existing networks of intermediaries and final beneficiaries. This enhances both the efficiency and quality of food service delivery within social assistance. Moreover, the approach is designed to be responsive to the evolving needs of vulnerable populations, ensuring sustained medium- to long-term support in changing social and economic conditions.

As rising social vulnerability places increasing pressure on public and private institutions, their ability to address complex social and food security challenges is being tested. Best practices highlight the crucial role of street-level adaptation, responsiveness, and collaboration between public actors and beneficiaries. These qualities are essential for effectively addressing both everyday needs and crisis situations. Moreover, they help safeguard local social welfare, promote stability, and integrate food assistance as a key complementary measure during and after disruptive events. Building on insights from best practices, one can identify key *governance mechanisms* that enhance the effectiveness and long-term sustainability of food security interventions at the local level.

Integrated Food Security Framework. A sequence of global-focusing events of different natures (climate and financial crises, pandemics, and wars) led European institutions to modify the food security agenda toward a more integrated policy framework. Best practices highlight the integration of food aid with other forms of material support, though the extent varies across projects. FEAD initiatives show that within a diverse mix of target populations and social measures, including food assistance, street-level bureaucrats (SLBs) operate in complex environments that require them to coordinate interventions across multiple services and sectors. This demands collaboration with both public and private actors, as well as the ability to integrate diverse knowledge and expertise. Embedding food assistance within a broader social inclusion framework reflects a holistic approach to addressing food insecurity as a dimension of social deprivation. This approach underscores the multi-functional role of SLBs, the value of cross-sector collaboration, and the need for policies that support integrated, tailored service delivery. By embracing these principles, policymakers can strengthen the role of SLBs in fostering comprehensive local welfare and social inclusion while also considering the broader environmental and social implications.

Collaborative Engagement Within Networks. Best practices regarding food aid and social inclusion share a common foundation of collaboration and trust-building within governance networks, including public institutions, private entities, and civil society organizations. The trust-building processes activated within the FEAD are both vertical—including multilevel organizations—and horizontal, across local entities. The emphasis on horizontal networking reflects strategic local efforts to ensure lasting impacts in combating poverty and social exclusion. However, the scale and composition of these networks vary significantly across different countries, influenced by their governance structures and levels of decentralization. In more decentralized systems, such as those in Germany, Italy, and Spain, local government officials, social workers, healthcare professionals, and educators play a central role in local food security initiatives. These professionals (involved in 39% of cases) integrate food assistance into existing local welfare policies, ensuring that aid is delivered in alignment with broader social support programs. The active participation of public-sector actors at the local level fosters policy socialization, allowing for greater adaptability and innovation in food aid strategies. This decentralized approach enables cities to develop context-specific interventions, drawing on the expertise of diverse actors while facilitating learning exchanges across different regions. However, differences can also be observed within the most decentralized countries; the absence of a unified national food security policy in the Federal Republic of Germany means that local initiatives are very active and highly fragmented.

Conversely, in more centralized systems—such as those in Portugal, Hungary, Ireland, Malta, Austria, Poland, the Czechia, and Finland—third-sector organizations and voluntary groups often operate alongside a narrower set of public institutions. In these countries, a high degree of delegation is given to intermediaries, such as food banks and charitable organizations, which take on much of the responsibility for food aid distribution and social support. This model relies on strong national oversight and policy coordination, with street-level bureaucrats (SLBs) who are primarily responsible for implementing predefined frameworks rather than shaping interventions at the local level. The varying degrees of public-sector involvement influence how FEAD-funded programs integrate food aid with broader social services. In decentralized settings, local actors have more flexibility to align food assistance with existing welfare programs, while in centralized systems, structured partnerships with third-sector organizations ensure efficiency in service delivery. Regardless of the governance model, the growing involvement of SLBs in food security networks highlights their crucial role in turbulent times in translating policy into practice. A key feature of the FEAD's evolving strategy is the acknowledgment of food aid not as a standalone measure but rather as part of a broader social inclusion framework. This shift aligns food security with other support mechanisms, thereby reinforcing the role of SLBs as agents of policy learning and institutional adaptation. The integration of food aid within public-private partnerships, particularly in response to the European poly-crises, underscores the increasing importance of holistic urban food welfare strategies. Understanding these dynamics is essential to design policies that strengthen the capacity of SLBs and enhance the effectiveness of food assistance programs within the wider social protection landscape.

Process-Oriented Tools for Collaboration. Evidence from the FEAD strategy across European countries highlights key innovations in street-level governance mechanisms. One of the most significant advancements is the introduction of a complex monitoring system based on the social vulnerability index, designed to ensure the equitable distribution of FEAD resources. Initially intended to help national decision-makers identify and prioritize vulnerable populations, this system has also facilitated the development of more collaborative processes and standardized procedural frameworks at the local level. The FEAD has played a crucial role in harmonizing the identification of intermediate and end beneficiaries across diverse institutional and local settings, thus contributing to reducing the discretion of SLBs. Additionally, the EU has enhanced accountability and feedback mechanisms by requiring detailed project reporting, ongoing monitoring of fund transfers, and periodic evaluations. These evaluations assess program effectiveness and financial allocations, providing insights for continuous improvement. Beyond financial support, the FEAD serves as a policy platform for knowledge exchange among national and local governments, practitioners, and stakeholders across Europe. It facilitates the honing of partnerships, the design of integrated food security programs, and the dissemination of best practices. The European Commission's networking efforts underscore the FEAD's dual role:

- **Substantive Integration**—Recognizing food aid as part of a broader approach to tackling social vulnerability.
- **Governance Collaboration**—Strengthening multi-actor partnerships that enhance collective learning and policy innovation.

The European platform has provided training, resources, and structured frameworks, enabling SLBs to operate within a widespread, holistic, collaborative, and integrated food assistance framework across Europe.

Street-Level Capacity to Learn from Beneficiaries. The best practice catalogue highlight that local street-level bureaucrats are increasingly oriented toward both intermediaries and end beneficiaries. However, there is considerable variation in the way different local systems define and address social vulnerability, influencing their inclusivity and target group selection. Some projects and Member States (MS) adopt a narrowly focused approach, targeting one or two specific vulnerable groups, such as people experiencing homelessness or the elderly. While effective in meeting the immediate needs of these populations, this approach risks overlooking the broader spectrum of social vulnerabilities within the community. Conversely, other projects and MSs embrace a more inclusive model, applying a broader definition of social vulnerability that encompasses diverse at-risk groups. These highly inclusive systems are better suited to addressing the complex and interconnected nature of social inequality. However, they also require more sophisticated coordination and resource management to function effectively. The FEAD network underscores the importance of local adaptation and flexible, context-sensitive approaches in food aid governance and social integration. By learning from beneficiaries and adapting to local needs, street-level initiatives can enhance their responsiveness, ensuring that food assistance programs remain both effective and inclusive.

That being said, by the end of 2020, European institutions acknowledged a significant shortfall in their goal of reducing the number of people at risk of poverty or social exclusion (AROPE) by at least 20 million. The actual reduction fell short by approximately 11 million. After eight years of FEAD initiatives addressing the effects of the poly-crises, long-term trends reveal stark contrasts among EU Member States (EU, 2024). Newer Member States, such as Poland, Czechia, and Croatia, have made significant progress. The proportion of people struggling to afford a proper meal declined from 18.5% in 2010 to 9.7% in 2022. In contrast, older Member States such as Germany, France, and Greece have seen minimal change, with figures remaining relatively stable—4.6% in 2010 and 4.9% in 2022. Among low-income households, newer Member States halved their rates from 38.5% in 2010 to 23% in 2022, whereas older Member States stagnated at around 13%. These disparities underscore the uneven effectiveness of measures to reduce food vulnerability across the EU.

In 2023, the gap between the general population and those at risk of poverty in terms of affording a proper meal across the EU stood at 12.8%. Hungary recorded the most significant disparity at 30.2%, followed by Slovakia (27.9%) and Greece (27.3%). Meanwhile, Sweden, Cyprus, Luxembourg, Portugal, and Ireland reported the smallest gaps, all below 4%.[5]

Looking ahead, the European Social Fund Plus (ESF+) for 2021–2027 assigns food aid a marginal role within the broader European social inclusion strategy. As food insecurity persists, national strategies—complemented by co-designed regional and local initiatives—are expected to play a central role in preventing severe material deprivation, mitigating food poverty, and fostering social inclusion.

Notes

1. The Material and/or Social Food Deprivation (DAMS) index is derived from 2 of the 13 indicators of Material and social deprivation (MSD) in the EU-SILC survey: the inability, due to economic constraints, to consume a complete meal every two days, and the inability to meet friends or relatives at least once a month to share a meal. This index estimates the proportion of individuals aged 16 and over experiencing at least one of these forms of deprivation, capturing both the material and social aspects of economic access to food.
2. Regulation (EU) No. 223/2014 of the European Parliament and of the council of March 11, 2014, on the Fund for European Aid to the Most Deprived.
3. As reported by Action Aid International, in recent years, several legislative efforts in the Italian Parliament have introduced universal access to school canteens, but these initiatives have faced obstacles. In April 2014, a bill titled 'Provisions to ensure equality in Minors' access to school canteen services' (AC2308) was introduced. The bill established equal access to canteens as a service upon individual request and as an essential standard by Article 117, paragraph 2, letter m, to ensure fair and consistent access across the entire national territory. However, this proposal should have been discussed in the parliamentary agenda. Then, in August 2015, Legislative Decree S2037 regarding 'Provisions on collective catering services' was presented. This decree sought to classify the school catering service as an essential public service. Despite being the subject of hearings in September 2016, it has yet to proceed to a vote.
4. The flexibilities introduced under Regulation (EU) No 2020/559 of April 23, 2020, which amended Regulation (EU) No 223/2014 as part of the CRII+ package.
5. EUROSTAT (2024). https://ec.europa.eu/eurostat/web/products-eurostat-news/w/ddn-20240712-1

References

Action Aid. (2024). *I numeri della povertà alimentare in Italia a partire dalle statistiche ufficiali*. Action Aid, Milano.

Adamczak-Retecka, M., & Śniadach, O. (2021). Towards a redefinition of the food security concept in the light of climate change: The European Union law perspective. *International Community Law Review, 23*(2–3), 283–293.

Alderman, H., Gentilini, U., & Yemtson, R. (2018). The evolution of food as social assistance: An overview. In H. Alderman, U. Gentilini, & R. Yemtson

(Eds.), *The 1.5 billion people question: Food voucher or cash transfer?* The Word Bank.

Ansell, C.K., Trondal, J., & Øgård, M. (2017). *Governance in turbulent times.* Oxford University Press. Retrieved December 31, 2023, from https://books.google.com/books?hl=it&lr=&id=gwuDDQAAQBAJ&oi=fnd&pg=PP1&dq=ansell+etal+2017+Governance+in+turbulent+times&ots=QtpPEcAEeA&sig=ob9NH46e5oH_9Sbx_j39y3V37ig

Bazerghi, C., McKay, F. H., & Dunn, M. (2016). The role of food banks in addressing food insecurity: A systematic review. *Journal of Community Health, 41*(4), 732–740. https://doi.org/10.1007/s10900-015-0147-5

Boyne, G. A., & Meier, K. J. (2009). Environmental turbulence, organizational stability, and public service performance. *Administration & Society, 40*(8), 799–824.

Busetti, S., & Righettini, M. S. (2023). Policy learning from crises: Lessons learned from the Italian food stamp programme. *Policy & Politics, 51*(1), 91–112.

Carlson, S. J., Andrews, M. S., & Bickel, G. W. (1999). Measuring food insecurity and hunger in the United States: Development of a national benchmark measure and prevalence estimates. *The Journal of Nutrition, 129*(2), 510S–516S.

Chaudhuri, S. (2003). *Assessing vulnerability to poverty: Concepts, empirical methods and illustrative examples* (p. 56). Department of Economics, Columbia University.

Cohesion Open Data Platform. (2021). *In profile: Fund for European aid to the most deprived (FEAD)*. Retrieved December 16, 2024, from https://cohesiondata.ec.europa.eu/stories/s/In-profile-Fund-for-European-Aid-to-the-Most-Depri/tdry-xg55

Dinan, S., Béland, D., & Howlett, M. (2024). How useful is the concept of polycrisis? Lessons from the development of the Canada emergency response benefit during the COVID-19 pandemic. *Policy Design and Practice, 7*, 1–12.

Dowler, E. A., & O'Connor, D. (2012). Rights-based approaches to addressing food poverty and food insecurity in Ireland and UK. *Social Science & Medicine, 74*(1), 44–51.

EC, DG Employment, Social Affairs and Inclusion. (2023). *Study supporting the preliminary evaluation of the support provided by ESF and FEAD under the Coronavirus response investment initiatives (CRII and CRII+): Final Report*, LUXEMBURG: Publications Office. Retrieved November 27, 2024, from https://data.europa.eu/doi/10.2767/160531

Ericksen, P. J. (2008). What is the vulnerability of a food system to global environmental change? *Ecology and Society, 13*(2), 18.

Esobi, I. C., Lasode, M. K., Anyanwu, C. I., Barriguete, M. F., Okorie, M. A., & Lasode, D. O. (2021). Food insecurity, social vulnerability, and the impact of COVID-19 on population dependent on public assistance/SNAP: A case study of South Carolina, USA. *Journal of Food Security, 9*(1), 8–18.

European Union. (2016). *Fund for European aid to the most deprived, FEAD case studies.* Publication Office of the European Union.

European Union. (2017). *Fund for European aid to the most deprived, FEAD case studies.* Publication Office of the European Union.

European Union. (2018). *Fund for European aid to the most deprived, FEAD case studies.* Publication Office of the European Union.

European Union. (2019). *Fund for European aid to the most deprived, FEAD case studies.* Publication Office of the European Union.

European Union. (2021a). *Diverse approaches to supporting Europe's most deprived-FEAD case studies.* Publication Office of the European Union.

European Union. (2021b). *Fund for European aid to the most deprived, FEAD case studies.* Publication Office of the European Union.

European Union. (2024). *2nd report on the state of food security in the EU (Spring 2024)*. DG Agriculture and Rural Development. Retrieved December 3, 2024, from https://agriculture.ec.europa.eu/document/download/a91b3841-6021-489e-b877-7f0f5278c88c_en?filename=efscm-assessment-spring-2024_en.pdf

FAO. (2023). *FAOSTAT: Suite of food security indicators.* FAO Knowledge Repository website. Retrieved December 16, 2024, from https://openknowledge.fao.org/server/api/core/bitstreams/f1ee0c49-04e7-43df-9b83-6820f4f37ca9/content/state-food-security-and-nutrition-2023/food-security-nutrition-indicators.html

FAO, IFAD, UNICEF, WFP and WHO. (2019). *The state of food security and nutrition in the world 2019: Safeguarding against economic slowdowns and downturns.* Roma, FAO Knowledge Repository website. Retrieved December 3, 2024, from https://openknowledge.fao.org/handle/20.500.14283/ca5162en

FAO, IFAD, UNICEF, WFP and WHO. (2023). *The state of food security and nutrition in the world 2023. Urbanization, agrifood systems transformation and healthy diets across the rural–urban continuum.* FAO Knowledge Repository website, Rome. Retrieved December 17, 2024, from https://openknowledge.fao.org/items/445c9d27-b396-4126-96c9-50b335364d01

Fondazione Brodolini, G. (2021). *Detailed report on the annual implementation reports for the operational programmes co-financed by FEAD in 2019.* Luxemburg Directorate-General Employment, Social Affairs and Inclusion (EU).

Galanakis, C. M. (2020). The food systems in the era of the coronavirus (COVID-19) pandemic crisis. *Food, 9*(4), 523.

Greiss, J., Schoneville, H., Adomavičienė, A., Baltutė, R., Bernat, A., Carrillo Álvarez, E., Delanghe, H., Goderis, B., Hermans, K., van der Horst, H., Michoń, P., Leite de Freitas Pereira, E. S., Correia Pereirinha, J. A., & Cantillon, B. (2022). *Food aid in Europe in Times of the COVID-19 crisis. An International survey project*. Working paper No.22/05, September 2022, CBS, Antwerpen University.

Guio, A. C., Marlier, É., & Nolan, B. (Eds.). (2021). *Improving the understanding of poverty and social exclusion in Europe*. Publications Office of the European Union.

Hermans, K., & Cantillon, B. (2025). How do European countries use EU-funded food aid and how important is it for the most deprived? JCMS. *Journal of Common Market Studies, 63*(1), 179–196.

Hermans, K., Greiss, J., Delanghe, H., & Cantillon, B. (2023). Delivering on the European pillar of social rights: Towards a needs-based distribution OF the European social funds? *Social Policy & Administration, 57*(4), 464–480.

Jochim, A. E., & May, P. J. (2010). Beyond subsystems: Policy regimes and governance. *Policy Studies Journal, 38*(2), 303–327.

Kumareswaran, K., & Jayasinghe, G. Y. (2022). Systematic review on ensuring the global food security and Covid-19 pandemic resilient food systems: Towards accomplishing sustainable development goals targets. *Discover Sustainability, 3*(1), 29. https://doi.org/10.1007/s43621-022-00096-5

Lipsky, M. (1980). *Street-level bureaucracy: Dilemmas of the individual in public service*. Russel Sage Foundation.

Loopstra, R., Reeves, A., & Stuckler, D. (2015). Rising food insecurity in Europe. *Lancet, 385*(9982), 2041.

López Cifuentes, M., & Fiala, V. (2022). Covid-19 as a chance for more food democracy in European cities? The responses of actors within Vienna's urban food system to the pandemic. *Cities, 131*, 104041. https://doi.org/10.1016/j.cities.2022.104041

Madama, I. (2018). Enhancing the visibility of social Europe: A practical agenda for 'the last mile'. In *Debating European citizenship* (pp. 261–266). Springer International Publishing.

Marsico, F., & Frigo, A. (2014). La Povertà Alimentare: Dal Pead al Fead, In Caritas Italiana website. *Il bilancio della crisi. Le politiche contro la povertà in Italia. Rapporto 2014.* Retrieved December 17, 2024, from http://s2ew.caritasitaliana.it/materiali/PDF/RapportoPolitichePoverta2014.pdf.

Mintzberg, H., & McHugh, A. (1985). Strategy formation in an adhocracy. *Administrative Science Quarterly, 30*, 160–197.

Moragues-Faus, A., & Battersby, J. (2021). Urban food policies for a sustainable and just future: Concepts and tools for a renewed agenda. *Food Policy, 103*, 102124.

Moret, W. (2014). *Vulnerability assessment methodologies: A review of the literature* (p. 360). FHI.

Morin, E. (1992). From the concept of system to the paradigm of complexity. *Journal of Social and Evolutionary Systems, 15*(4), 371–385. Retrieved

December 27, 2023, from https://www.sciencedirect.com/science/article/pii/1061736192900248

Morin, E. (2017). La sfida della complessità. *Le Lettere.*

Nordensvärd, J., Li, B., Turcu, C., Qian, J., Byun, Y. H., Li, Y., Sommar, C. J., & Rotolo, M. (2023). State and communities in urban food governance: Lessons from COVID and insights for the future. *Urban Governance, 3*(2), 93.

O'Connor, N., Farag, K., & Baines, R. (2016). What is food poverty? A conceptual framework. *British Food Journal, 118*(2), 429–449. https://doi.org/10.1108/BFJ-06-2015-0222

Oldroyd, L., Eskandari, F., Pratt, C., & Lake, A. A. (2022). The nutritional quality of food parcels provided by food banks and the effectiveness of food banks at reducing food insecurity in developed countries: A mixed-method systematic review. *Journal of Human Nutrition and Dietetics, 35*(6), 1202–1229. https://doi.org/10.1111/jhn.12994

Pérez-Escamilla, R., Gubert, M. B., Rogers, B., & Hromi-Fiedler, A. (2017). Food security measurement and governance: Assessment of the usefulness of diverse food insecurity indicators for policy makers. *Global Food Security, 14*, 96–104.

Sen, A. (1982). The food problem: Theory and policy. *Third World Quarterly, 4*(3), 447–459.

Simmet, A., & Stroebele-Benschop, N. (2023). Food bank operations during the COVID-19 pandemic in Germany. *Journal of Hunger & Environmental Nutrition, 18*(3), 356–371.

Tashtoush, F. M., Al-Zubari, W. K., & Shah, A. (2019). A review of the water–energy–food nexus measurement and management approach. *International Journal of Energy and Water Resources, 3*, 361–374.

WBAE. (2023). *Scientific advisory board.* Department of Agricultural Policy, Food and Consumer Health Protection, Berlin.

CHAPTER 3

Lessons from Managing the Food Aid Program During the Pandemic: Insights from Italy

In Italy, the COVID-19 pandemic has reshaped national discussions on food security policies, prompting their redesign and integration into a broader social policy framework. As one of the EU Member States that was most affected by the pandemic, Italy serves as a key example of how food aid programs have been developed, adopted, and implemented at both national and local levels in response to the crisis.

Food poverty has long been a concern in Italy, dating back to its transition from an agrarian to an industrial society. Historically, over half of Italian households faced food insecurity (Fiocco, 2004). Today, Italy is one of Europe's most developed countries, with a robust economy ranking among the top in the EU and a GDP of approximately $2 trillion, making it the third-largest economy in the Eurozone. However, food poverty has been on the rise in recent decades. According to recent statistics, around 12% of the Italian population—approximately six million people—still live in conditions of food poverty (Action Aid, 2024). Although the percentage of individuals at risk of poverty or social exclusion in Italy declined between 2016 and 2018, it remained significantly higher than the EU average during the period from 2014 to 2018. In 2019, 4.6 million households in Italy were still living in absolute poverty, with 40.5% residing in northern regions and 45.1% in the south of the country. Among them, one in four were minors (1.14 million people), while nearly one in three were foreigners (1.4 million) (ActionAid, 2020). To address these

M. S. Righettini, *The Governance of Food Security in Europe*,
https://doi.org/10.1007/978-3-031-98083-1_3

challenges, the Program for European Aid to the Deprived (PEAD), active from 1987 to 2013, and the European Fund for Aid to the Most Deprived (FEAD), active between 2014 and 2021, played a crucial role in combating food and social poverty in Italy. PEAD, in particular, was a vital resource for charitable organizations across EU member states, providing food for free distribution to those in need.

In Italy, the PEAD Fund was administered by the Agency for Agricultural Payments (AGEA) on behalf of the Ministry of Agricultural, Food, and Forestry Policies. AGEA managed the fund through a system of procurement tenders, overseeing all stages of production and delivery. This work was carried out in partnership with seven formally recognized charitable organizations—Associazione Banco Alimentare di Roma Onlus, Associazione Sempre Insieme per la Pace, Caritas Italiana, Comunità di Sant'Egidio, Croce Rossa Italiana, Fondazione Banco Alimentare Onlus, and Fondazione Banco delle Opere di Carità—all registered in AGEA's official register. By establishing a cohesive, subsidiary, and extensive public–private network, these seven organizations have distributed food products to over 15,000 local charitable organizations, delivering them directly to those in need.[1] Following the end of the PEAD and before new European decisions were made, the Monti Government established the National Fund for the Distribution of Food to People in Need with Law 134/2012. This fund was designed to be supported by public and private resources to increase the volume and variety of food that charitable organizations provide. Amid the operational challenges of implementing the National Fund and adapting to new FEAD regulations, the Renzi Government initiated an experimental program: the Social Card.

Initially introduced in 2008 (Berlusconi Government), the first version of the Social Card was intended for seniors over 65 and children under 3 in low-income households, providing 40 euros per month to purchase food and pay utility bills. With the FEAD's support, the updated Social Card became part of a broader national anti-poverty strategy, moving beyond essential food assistance to include a focus on social inclusion and integration. Across the country, FEAD has mobilized vital resources to support the most vulnerable populations, fostering collaboration between local and national authorities while promoting public–private partnerships to enhance food access. This approach has enabled Italy to integrate local welfare policies with concrete measures to reduce poverty, particularly addressing the food needs of those most at risk of social exclusion. National policies encouraging food donations have further driven many households to seek assistance from charitable organizations and FEAD partners, such as Food Banks, Caritas, the Red Cross, and numerous smaller local groups. According to the Ministry of Social Policies and Labor,[2] the number of people receiving FEAD food aid rose significantly owing to the economic downturn during and after the pandemic. Beneficiaries increased from 2.1 million in 2019 to nearly 3 million in 2021, before slightly declining to just over 2.8 million in 2022. The proportion of women, minors under 16, and individuals over 65 has increased, while the share of migrants, after declining in 2021, has returned to 2019 levels (Action Aid, 2024). Naples (243,000 beneficiaries), Milan (214,000), and Rome (197,000) were the metropolitan cities with the highest number of FEAD beneficiaries.

The evolution of food security governance in Italy is an interesting case. COVID-19 marked a transition from a program entrusted to charitable organizations, where food aid had been treated as an isolated component in urban governance (Bandera et al., 2016), to a more integrated public–private approach. The pre-pandemic approach has reinforced the critical role of private (economic) and nonprofit organizations in bridging gaps in public support and meeting immediate needs. In Italy, food security has typically been seen as a peripheral aspect of social policies and local welfare, devolved to nonprofit and civil society organizations, strictly linked to the food waste and surplus management and donation collection and redistribution (Busetti, 2019). The approval of the national reform through the 'Good Samaritan' law (2003) 'Regulation of the distribution of food products for social solidarity purposes' and the 'Gadda' law (2016) 'Provisions concerning the donation and distribution of food and pharmaceutical products for social solidarity purposes and to limit waste,' aimed at combating food waste and loss while incentivizing donations. Before 2016, food donors in Italy faced heavy fiscal burdens. However, the 2016 law significantly simplified the process, improving communication and expanding the range of goods eligible for donation. By reducing bureaucratic hurdles, the reform encouraged food businesses to donate more freely, thus increasing overall contributions (Busetti, 2019). The pioneering reform strengthened the charitable approach to food security and emphasized the alliance between large-scale retail trade (GDO), food banks, and third-sector organizations. However, it also limited the role of local public actors and municipalities, preventing them from implementing proactive programs and leaving them largely inactive until the onset of the COVID-19 crisis. As a result, efforts to develop and implement food security interventions relied on existing, regionally varied social capital. Since food banks, humanitarian organizations, and volunteers are not evenly distributed across the country, this has led to geographical disparities in providing essential urban food policies. Moreover, the lack of recognition of universal access to primary school canteens in Italy has significantly contributed to increased food insecurity among children, particularly those from low-income households. Without access to nutritious meals provided during the school day, these children are more vulnerable to poor nutrition, which can negatively impact their health, development, and academic performance. For many families struggling with financial hardship, school meals are a crucial source of balanced nutrition for their children. When access to these meals is restricted or unavailable, it deepens the cycle of food insecurity, exacerbating the challenges faced by disadvantaged communities and growing social inequalities. In areas where school canteens are not available, children take advantage of the opportunity to have a balanced meal, which can impact their health and school performance. European institutions have repeatedly acknowledged this gap in local food services, aiming, in addition to promoting interventions, to support fairer policies that guarantee all children access to nutritious meals at school, regardless of their socio-economic background.

The crisis related to the pandemic of 2020 and 2021 further exacerbated a new wave of poverty and vulnerability, increasing, on the one hand, the demand for food aid and the proportion of individuals receiving support from civil society organizations, such as Caritas and Food Banks, and, on the other hand, increasing the demand for public interventions and better public–private coordination in food assistance (ISTAT, 2021). In 2021, about 5.6 million people, including nearly one-third of all non-national residents, were unable to buy goods and services essential to achieve a minimally acceptable standard of living as defined by the Italian government.[3] An estimated 22% of people living in Italy are still at risk of food poverty or are food insecure (ISTAT, 2024).

The adoption of the National Food Assistance Program (FAP) in early 2020 aimed to combat food insecurity and vulnerability amid the escalating health and economic crisis caused by COVID-19, accelerating its implementation at the municipal level. For the first time, all municipal street-level bureaucrats (SLBs) were tasked with adapting the national food emergency program to local needs, thus marking a significant shift in governance. This process underscored the growing role of decentralized decision-making in crisis response, enabling municipalities to tailor interventions based on regional socio-economic conditions and food insecurity levels. The flexibility granted to SLBs improved policy responsiveness and fostered stronger collaboration between local governments, nonprofit organizations, and food distribution networks. However, it also exposed disparities in administrative capacity, as resource-rich municipalities implemented more effective measures than those with limited institutional support.

The following sections examine food aid programs in 99 Italian provincial capital cities, focusing on the design and distribution of food vouchers and the changes observed between the first and second waves of the pandemic. Using the analytical tool of CM (Table 1.2), this research explores how SLBs implemented the national food security program during the crisis, whether and how they established collaborative governance mechanisms with beneficiaries, and what lessons emerged from the voucher distribution system.

3.1 Crisis Response in Action: Street-Level Challenges in Managing Food Aid

The response to the crisis and emergency brought about by the pandemic in Italy has predominantly been analyzed through the lens of national policymaking, focusing on the actions taken by state and regional governments, particularly within the healthcare system (Capano et al., 2020; Capano & Lippi, 2021; Malandrino & Demichelis, 2020). However, more attention ought to be paid to local initiatives to alleviate the pandemic's negative social impacts and address food emergencies (Gofen & Lotta, 2021). The pandemic created a unique intersection where various program streams, local challenges, and solutions converged—some of which were already recognized, while others, previously overlooked or neglected, gained newfound public attention in the context of the crisis (Kingdon & Stano, 1984). The shift underscores the need to analyze multilevel responses and redundancies in tackling the broader social challenges that were intensified by the pandemic. Specifically, this chapter focuses on the mechanisms guiding SLBs' implementation and design of the take-up process for end beneficiaries of a national program. It involves the continuous adjustment of screening (identifying groups that are potentially at risk), targeting (selecting the most suitable beneficiaries), and delivery (ensuring effective benefit distribution). While a well-functioning take-up system reflects program administrators' efforts to enhance managerial capacity and program effectiveness (Kleven & Kopczuk, 2011), these measures can inadvertently increase the application burden for citizens, potentially leading to non-participation among eligible and more vulnerable individuals (Busetti, 2019). Therefore, one reason social food programs often adopt a 'universal' approach is to reduce and simplify selection and access criteria (Righettini & Sbalchiero, 2017). In times of emergency and crisis, the time dimension is crucial (Howlett, 2019), and a top-down implementation approach is often inadequate (Elmore, 1979). The Italian FAP is a case of multi-agency collaboration strategy, fostering horizontal lines of cooperation (Klijn & Koppenjan, 2016), and teamwork of street-level bureaucrats (SLBs) (Emerson et al., 2012). Successful implementation during crises follows a backward-mapping logic (Elmore, 1979), where SLBs proceed through trial and error, continuously redesigning the implementation process. Nevertheless, the take-up systems at the street level can reveal specific gaps in policy target recognition and the need to learn how to fix existing procedures rapidly. The presence of imperfect information about the target population can jeopardize a program's equity (Sohn & Kwon, 2021). For example, outdated criteria and legislation that fail to match rapidly evolving situations (Currie, 2004; Davidovitz et al., 2021; Hupe & Buffat, 2014; Sager et al., 2020) can make it difficult to assess individuals' or households' income-generating capacities accurately. This is especially true for food aid, where social stigma intersects with the challenge of understanding food vulnerability in its complex dimensions, thus affecting individuals and households (Action Aid, 2023, 2024).

In FAP programs, local bureaucrats can make two main types of errors: the first related to beneficiary identification and social benefits allocation (Kleven & Kopczuk, 2011); the second is a classification error (false negatives/false rejections) and assessment errors (false positives/false awards). The first case involves the failure to enroll in social assistance programs, which can result from errors in how public offices determine the criteria for awarding benefits and social protection standards. The second type pertains to incorrect attributions. There can be a close correlation between the two types of errors because the existence of the first (non-registration) often leads to the second, which is the incorrect allocation of benefits. Economic literature explains these errors with three main causes: welfare stigma, transaction costs, and imperfect information. Some individuals may refuse to access benefits because they consider food aid socially stigmatizing. In contrast, others do not access them owing to the high transaction costs associated with demand and access procedures, resulting from detailed eligibility criteria, stringent documentation requirements, complex and time-consuming forms, or requiring interviews and tests. Finally, a third category of individuals is often excluded from programs owing to information asymmetries, exacerbated by the frequent recertification processes required to continue receiving benefits in prolonged crises. This can result in applicants being denied assistance because they must meet new administrative requirements within the specified timeframe. Enhancing and intensifying interactions and collaboration among various stakeholders is essential to avoid asymmetries, working toward common objectives through shared strategies (Lodge & Wegrich, 2014). Adopting a multi-agency approach helps achieve this, through the use of procedural tools such as networking, teamwork, and the alignment of goals (Abramson & Mittu, 2006; Scerri et al., 2006; Scott & Gong, 2021). What matters, ultimately, is that collaborative procedures enable program implementers to react promptly to changes in the needs of end beneficiaries. The intensity of the pandemic crisis prompted varying degrees of interaction and stakeholder collaboration, aiming to reduce information asymmetries in screening and targeting beneficiaries and delivering essential benefits. Interviews with street-level operators confirm that external collaborations with NGOs act as strategies of institutional adaptation, enabling quicker and more effective outreach to beneficiaries, especially in conditions of unknown complexity and uncertainty (Ansell et al., 2021).

In ordinary times, the limited rationality of profiling, monitoring, and selection activities introduces cognitive asymmetries into the application process, which can hinder the participation of those not in poverty yet but most vulnerable. Empirical research on FEAD management has shown that the complexity of procedures and administrative issues reduces enrollment in social programs and, thus, encourages classification errors and bias (Currie, 2004; Kleven & Kopczuk, 2011; Van Oorschot, 1991). One source of assessment errors can be the failure to adjust public databases, especially local ones, and the need for more intervention of social services, which continue to operate based on outdated or insufficient knowledge in rapidly changing contexts. Since acute crises produce rapid and unexpected changes but require equally rapid interventions, using ready-made solutions often has an intrinsic limitation in grasping new needs and targeting potential new crisis policy recipients (Ansell et al., 2020). In times of crisis and emergency, such as the pandemic, also the income criterion can be inadequate for two fundamental reasons: many incomes are not officially declared, and a temporary reduction in income does not necessarily constitute a situation of social insecurity, poverty, or structural destitution but rather vulnerability linked to rapid and unpredictable contextual changes. Attention to the design of monitoring systems and eligibility criteria for access to public support allows for 'careful beneficiary identification' (Barzelay, 1992; Barzelay & Thompson, 2010). This seemingly minor aspect in implementation studies is increasingly recognized as a central problem in targeting policies (Knill & Tosun, 2020).

The following sections provide a deeper analysis of the management of food vouchers across 99 Italian provincial capital cities, with a particular focus on the actions taken by municipal street-level bureaucracies (SLBs) to address the emergency and food security needs of individuals and households. This work's analysis includes data from municipal deliberations and employs a diachronic comparison between the management of the first and second rounds of food voucher distribution during the two pandemic waves, along with a synchronous comparison across cities in these respective phases.

The examination of the National Food Security Program's implementation from April 2020 to April 2021, within the context of the COVID-19 emergency, reveals remarkable changes in governance mechanisms for the Food Assistance Program (FAP) (Ansell & Gash, 2008; Ansell, 2012; Ansell et al., 2020; Emerson et al., 2012; Emerson & Nabatchi, 2015a, 2015b). Specifically, one must note the multilevel decentralized governance structure, the challenges related to the local implementation of the FAP, and the bureaucratic and administrative barriers encountered. This work also explores whether and how procedural flexibility has been implemented in food vulnerability recognition and assistance; how SLBs have leveraged and diversified public–private collaborative networks; and how these collaborations have been institutionalized. Additionally, the work examines how monitoring tools have supported adaptation and flexibility in the implementation process.

3.2 The National Food Aid Program: Design and Framework

The virus was first identified in Italy on January 31st, 2020, and the Italian government declared a state of emergency in February. Eleven municipalities in northern Italy, in Lombardy and Veneto, were identified as the centers of the two largest Italian clusters and therefore isolated. By early March 2020, the virus had spread across the entire country, affecting, in particular, the northern regions. On March 8th, 2020, Prime Minister Giuseppe Conte extended the quarantine to the entire region of Lombardy and 14 other northern provinces and the next day to the entire country, putting over 60 million people in quarantine. The Italian Prime Minister subsequently ordered the closure of almost all commercial activities, except for supermarkets and pharmacies, and on March 21st, he closed all non-essential businesses and industries, further restricting people's movement. Alarming economic concerns thus compounded health issues, as the lockdown led to a sharp decline in productive activities, resulting in a GDP reduction of over 8%. The loss of jobs, amounting to 600,000 in 2020, exacerbated these challenges. Meanwhile, an additional issue arose in cities, where panic-buying at supermarkets for essential goods became a major concern, which the central government attempted to address. Ordinance No. 658 of March 29, 2020 of the national Chief of the Civil Protection Department, concerning 'Further Urgent Civil Defense Measures Related to the Health Risk Associated with the Onset of Viral Communicable Diseases,' introduced food vouchers as a distribution tool available to municipalities.[4] The ordinance allocated additional funds worth 400 million euros to municipalities. Eighty percent of the total sum was distributed based on population numbers. In contrast, the remaining 20% was distributed based on the 'distance between the per-capita income of each municipality and the national average' (Art. 2). The pandemic was an unforeseen event that brought to light a widespread sense of food insecurity across the population and exposed the often-overlooked issue of food vulnerability within urban environments.

The National Food Aid Emergency Act mandated that municipalities across all levels implement a food aid program employing government-provided resources. This directive led to an unexpected, major shift in municipal agendas nationwide. For the first time, every municipality was required to prioritize food aid. The program required municipalities to establish and engage a network of stakeholders, including public entities and primarily private third-sector organizations, responsible for managing meal vouchers. This collaborative approach aimed to ensure that the most vulnerable populations were accurately identified and that their access to essential benefits was secured.

In Italy, food security measures are generally overseen by the social services offices within each municipality, with few exceptions. The decree explicitly assigned these offices the responsibility of identifying the households that were most vulnerable to the economic impacts of the pandemic and those in need of food aid. This directive assigned local social services the task of accurately identifying and supporting food assistance beneficiaries during this critical period. This process also required municipalities to determine the appropriate level of assistance to address the most urgent and essential needs, primarily focusing on individuals who had not yet received other forms of public support, such as basic income. To streamline this, the decree simplified procedures, allowing municipalities to distribute food vouchers for purchasing food and other essential products from businesses listed on the municipality's official website. The government's ordinance institutionalized a collaborative approach to food security governance by enabling municipalities to partner up with third-sector organizations to purchase and distribute these goods. Municipalities were encouraged to collaborate with entities that were already active in food distribution through the European Aid to the Most Deprived (FEAD) Program since 2014. To support this, the Ministry of Labor and Social Policies provided a region-specific list of FEAD leading organizations on its website. Additionally, third-sector personnel and volunteers involved in food distribution were exempt from travel restrictions, thus ensuring timely delivery of aid.

Given the novelty of this provision, the National Association of Italian Municipalities (ANCI) issued general guidelines to help standardize and facilitate the interventions across regions and administrative capacities. Despite the complexity of the emergency and the challenges posed by remote work for most administrative staff, the rapid response by municipal administrations highlighted both the limitations and potential of local governance in addressing food security during a crisis. Figure 3.1

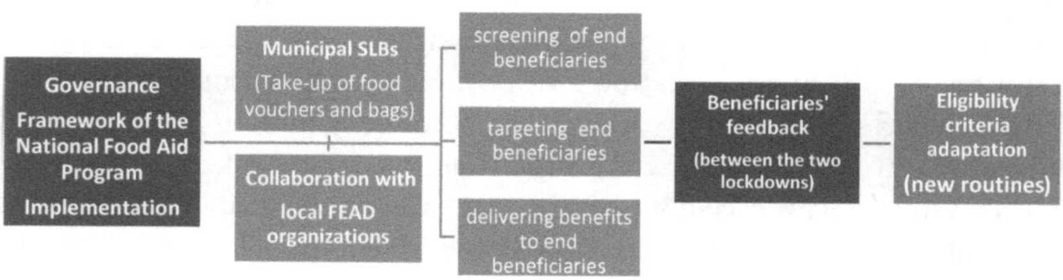

Fig. 3.1 The multilevel governance framework of the Italian national food aid program during COVID-19. (Source: Made by the author based on official documents)

illustrates the common governance framework for the management of the take-up scheme. It involves public and private urban street-level operators in implementing the National FAP designed to combat food insecurity during the two waves of the pandemic.

The design of take-up systems is crucial in reflecting the capacity of local street-level officials to identify and reach the policy target during a crisis (Lodge, 2015; Lodge & Wegrich, 2014). These systems are particularly significant during emergencies, such as the COVID-19 pandemic, as they determine how effectively benefits can be distributed to those in need. The 'take-up system' refers to the entire process involving screening, targeting, and delivery, which allows local street-level bureaucracies (SLBs) and FEAD partners to identify and reach the ultimate beneficiaries. This system aims to ensure a proper match between the demand for and supply of benefits, enabling local authorities to meet the urgent needs of vulnerable populations (Kleven & Kopczuk, 2011). At the core of this system are street-level workers, who, as identified by the national FAP, include municipal social workers, as well as operators from third-sector and charitable organizations. These groups have long served as an essential interface between local public administrations and existing multilevel food security programs (Thomann, 2015; Thomann et al., 2018). The relationship between the take-up system and the governance changes within the FAP during crises highlights the adaptability and responsiveness of local systems in the face of rapidly changing needs. However, these governance changes also resulted in challenges in ensuring consistency and equity across different regions. Variations in administrative capacity, local resources, and pre-existing networks of food security actors impacted how effectively local authorities could implement the FAP and manage the take-up system. The take-up system's role in the broader governance change is also intertwined with the shift toward more collaborative, public–private partnerships. The increased reliance on these networks also helped address some of the challenges related to targeting and delivery, as third-sector organizations often had closer ties to local communities and were better positioned to reach those most in need. In summary, the relation between the take-up system and FAP governance changes during crises reveals how local street-level bureaucracies adapt to shifting circumstances, balancing the need for rapid response with the challenges of ensuring equity and consistency.

3.3 FAP Knowledge Management from the Street Level: Weaknesses and Strengths

During the documentation-gathering process for the initial implementation of the national Food Aid Program (FAP) across the 99 Italian provincial capitals, a series of in-depth interviews were conducted in four municipalities in the Veneto region. These interviews aimed to provide a detailed understanding of the on-the-ground situations and gather feedback from within the administration regarding the newly introduced procedures. These procedures are part of a broader knowledge management cycle, encompassing local knowledge acquisition, translation, and spreading stages. The municipal operators to interview were chosen based on several shared factors: (a) they were in geographical areas that were heavily affected by the pandemic during its initial phase; (b) they had not previously adopted a specific food policy, including the security dimension; and (c) had nonetheless participated in food assistance programs in partnership with humanitarian and third-sector organizations as a means of income support (Busetti & Righettini, 2023). The 12 in-depth interviews conducted between the two pandemic lockdowns (April and June 2020) across four cities—Venice, Padua, Verona, and Vicenza—provided crucial insights into the evolving food aid landscape during the first phase of the COVID-19 pandemic. The interviews revealed an unexpected surge in food aid applications, driven by the emergence of a previously 'unknown' population segment—an estimated 20–30% more individuals than those already served by municipal services. This new demographic mainly consisted of immigrants who engaged in undeclared work and precarious workers in the tourism and food distribution sectors, who were particularly vulnerable owing to food and housing insecurities. Identifying this group was a key learning experience for local governments. As a result, plans based on previous data proved inadequate, and the initial financial resources were insufficient to meet the increased demand.

In some cases, the number of applications doubled the available funds (Interviews 1 and 3). This situation prompted local governments and social workers to request additional funding in anticipation of a second wave of the pandemic. However, between the two lockdowns, the context changed once again: the availability of vaccines, new social distancing rules, and the shift to remote work led to a decrease in the number of people seeking food aid. As a result, the additional financial resources surpassed the actual demand.

The mismatch between the demand for and supply of food aid in turbulent times suggested that street-level bureaucrats (social workers) were unprepared to deal with fluctuating needs and needed to be more agile in reallocating resources as the situation evolved. This underscored the importance of establishing robust communication channels, and real-time data collection became crucial to better predict and respond to future crises (Interviews 1, 2, and 3). Additionally, the street-level bureaucrats who were interviewed emphasized the need for local contingency funds and improved coordination with non-governmental organizations to address emerging needs more swiftly. This experience highlighted the need for flexible tools, adaptable strategies, and the ability of frontline workers to identify and respond to rapidly changing community needs.

The in-depth interviews highlighted the key challenges faced by municipal SLBs in implementing the FAP program, particularly in identifying and reaching beneficiaries during crises. Table 3.1 summarizes these challenges, especially in managing information flows about food vouchers during the first wave of the pandemic, and outlines strategies for improving these processes in future crises.

The challenges faced by Italian municipal operators during the first wave of the pandemic mirror those identified in the pre-pandemic Eurocities study, which revealed common issues among street-level bureaucrats (SLBs) across European cities (Eurocity., 2021). The implementation of food security programs exposed the complexity of managing a constantly shifting target population, further complicated by ongoing crises. This situation underscored the importance of collaborating with external partners to effectively manage knowledge, identify hidden vulnerabilities, and allocate resources for those in need.

3.4 Adapting to Change: The Evolution of Municipal Take-Up Processes

As 'frontline work is an expression of embedded agency and pragmatic practices rather than discretionary decision-making' (Lotta et al., 2023, p. 14), the evolution of national FAP implementation and management at the municipal level was closely monitored between the first and second phases of the pandemic to fully assess SLBs' capacity to adapt in times of crisis.

Table 3.1 Learning challenges identified by the Italian municipal officials and operators during the first COVID-19 wave and FAP management

Learning challenges in FAP management at the city level during the first wave of the pandemic crisis (2020)	*Expected improvements in street-level knowledge acquisition, translation, and diffusion system (lessons learned)*
Outdated local databases of social services (knowledge acquisition)	Backward mapping by updating monitoring and adjusting the beneficiary profiling, therefore considering previously undervalued and interconnected needs traditionally treated separately, such as food assistance, housing, and type of employment
The gap between food assistance supply and demand (knowledge acquisition)	Social target construction in turbulent times revealed new dimensions of social fragility that must be investigated and certified at the local level
Limited data flow integration and digitization (knowledge translation)	Digital innovation in integrating and digitizing data flows on beneficiaries across sectors has proven strategic in accelerating and making monitoring, beneficiary profiling, and policies more reliable and effective
Inadequate digital tools to deliver benefits (knowledge translation)	Digital innovation in the interface with the ultimate beneficiary must become routine on the application (demand) and voucher delivery (supply) sides. It ensures safer and easier access to beneficiaries and may be helpful in future crises
Red tape (knowledge translation)	Streamlining and simplifying administrative procedures improve program timing and delivery capacity (reducing entry checks and eliminating formalities)
Inadequacy of external collaboration (knowledge translation)	Strengthening partnerships and trust-based interactions with third-sector organizations, volunteers, and some national organizations that frequently deal with poverty and food insecurity (Red Cross, Caritas, Food banks)
Silo mentality (knowledge diffusion)	The integration of the operations of various municipal offices (social, ICT, registry, economic activities, construction) and adequate management of the emergency policy process rely on reducing or even eliminating the administrative silo mentality and the rigid division of duties and responsibilities during turbulent times
Resource sustainability (knowledge diffusion)	The temporary and exceptional nature of crisis funding underscores the need to transition from emergency responses to sustainable, long-term strategies for addressing growing food poverty and social vulnerability

Source: Made by the author based on interview content

In autumn 2020, Italy's Ordinance No. 658 of 2020 and Decree-Law No. 154 'Ristori ter' of November 23 allocated an additional 400 million euros to municipalities for a second wave of food voucher distribution. Data collected from 99 out of 107 Italian provincial capitals highlighted changes in food aid take-up systems between the first and second pandemic waves. The local implementation of the national program was monitored in real time from April 2020 to April 2021, tracking municipal resolutions that authorized the distribution of financial resources twice. This systematic approach captured the dynamic adjustments and adaptations in FAP implementation between the two distribution phases. The data provided a detailed account of how municipalities responded to shifting community needs during the pandemic.

The research examines the managerial aspects of the FAP implementation, drawing on insights from interviews with municipal officials and key stakeholders. These interviews were instrumental in identifying critical challenges in the FAP take-up process, including:

I. **Screening and targeting the population**: defining and refining eligibility criteria to adapt to the evolving socio-economic landscape.
II. **Delivering food vouchers**: transitioning to digital platforms for applications and assessing their impact on accessibility, efficiency, and overall program management.
III. **Collaboration with external organizations**: partnering up with FEAD nonprofits, charities, and community groups to enhance outreach and ensure food aid reaches vulnerable populations.

By examining these key aspects, the research offers a comprehensive view of how local governance navigated the complexities of collaborative learning in the management of food assistance programs during an unprecedented crisis.

Screening and Targeting Population

Recognizing food aid beneficiaries is a key component of the local learning process. Therefore, the first focus is on how eligibility criteria for FAP beneficiaries evolved, highlighting the crucial role of street-level bureaucrats (SLBs) in screening and identifying those in need. Table 3.2 outlines the key eligibility criteria adopted by municipalities and the percentage of

Table 3.2 Net change in beneficiary eligibility criteria between the first and second food voucher delivery

Eligibility criteria	*Admission of non-residents*		*Income threshold*		*Asset values*		*Extra bonus for special categories*	
Total 99 municipalities	N	%	N	%	N	%	N	%
Number of municipalities changing eligibility criteria between the first and second food voucher delivery	23	23.2	38	38.3	36	36.3	19	19.9

Source: Made by the author based on the original database 'Alimentaria'

those that revised them between the first and second waves of the pandemic. The criteria included place of residence, income thresholds, asset values, and identifying vulnerable groups such as disabled individuals, infants, and the elderly.

Initially, eligibility was primarily based on place of residence and income. However, many municipalities lacked precise income thresholds, creating confusion for applicants who fell outside loosely defined parameters. This issue was compounded by a reliance on outdated 2018 tax returns, which excluded individuals who lost their jobs in 2019 but exceeded income limits based on their previous earnings. As a result, many vulnerable individuals were ineligible for aid despite not qualifying for other support programs.

The rigidity of the initial criteria, combined with the evolving economic crisis, highlighted the need for a more flexible and responsive eligibility framework. Over time, some municipalities adjusted their criteria to reflect current socio-economic conditions, incorporating recent income data and recognizing the vulnerabilities of groups such as migrants and single-parent families. Interviews with street-level bureaucrats (SLBs) indicated that charitable and third-sector organizations played a crucial role in refining beneficiary screening by identifying at-risk populations.

The adjustments of the eligibility criteria underscored the importance of adaptability in social assistance programs, particularly during crises, to prevent inadequate interventions. Given the resource constraints of public agencies, collaboration with intermediate beneficiaries and continuous feedback collection were essential to improve the programs. SLBs also stressed the need for better data collection and inter-municipal cooperation to ensure a more cohesive and equitable aid distribution system. While most municipalities operated independently in defining FAP eligibility, some coordination efforts emerged. In regions such as Friuli-Venezia Giulia and the Autonomous Province of Trento, social services were managed at the provincial rather than municipal level, thus facilitating the harmonization of application procedures, eligibility criteria, and voucher distribution.

Twenty-three municipalities adjusted their place of residence criteria to include non-residents, following a ruling by the Regional Administrative Court of Lazio, which recognized the universal nature of the food assistance program. Additionally, 38 municipalities modified income thresholds and 36 adjusted the asset value threshold. Nineteen municipalities also revised the extra bonuses related to the presence of vulnerable individuals within households. These discrepancies in eligibility criteria and reliance on outdated income data underscore the urgent need for social services to adopt more flexible and responsive systems. One key implication is integrating real-time data, allowing for more accurate assessments of current financial situations. Moreover, social services must establish clear and precise eligibility criteria to minimize confusion and ensure fair access to aid. The geographical variance in adopting digitization and other changes is also noteworthy. Cities in the northwest of Italy, for example, introduced fewer modifications to income criteria. This may suggest that food insecurity in these northern cities exhibited more stable characteristics between the first and second waves of the pandemic. The relative stability could be attributed to pre-existing digital capabilities in many provincial capital cities in the north, which provided a more robust foundation to manage food aid in a timely manner during the crisis.

The screening and targeting experience highlighted the need for continuous learning and adaptation within social services. Regularly updating policies and criteria based on emerging needs and feedback from beneficiaries can help prevent future exclusions. Furthermore, social services ought to consider building more robust support networks and fostering collaborations with other agencies to provide comprehensive aid to those most affected by crises. This ongoing learning process is crucial for enhancing the effectiveness and equity of aid distribution in future emergencies. By embracing flexibility, clear communication, and interagency cooperation, social services can better respond to the dynamic needs of vulnerable populations during times of crisis.

Procedural Innovation in Reaching the Social Target: Digitizing Food Vouchers' Application and Delivery Processes

The digitization of food voucher applications and distribution emerged as a key innovation in pandemic-era food assistance programs (Galanakis et al., 2021). By integrating technology with community support systems, these advancements streamlined aid delivery, improving efficiency, accessibility, and responsiveness to vulnerable populations that were 'left behind' (Giest, 2025). Between the first and second waves of the pandemic, 47.7% of provincial capital cities modified both the voucher format and distribution process, shifting from paper-based to digital systems (Table 3.3). This transition reflected a broader move toward digitization, thus enhancing efficiency and accessibility for beneficiaries.

This shift toward digitization during the pandemic presented both challenges and opportunities for municipalities. The need for safer distribution methods and more efficient beneficiary verification drove the adoption of digital solutions. Digital vouchers streamlined aid distribution by reducing physical contact and enabling faster, more accurate eligibility assessments. The widespread adoption of remote work further supported this transition, with municipal information systems managers playing a key role in implementing new digital systems. However, not all municipalities were equally prepared for rapid digitization. Those lacking infrastructure or expertise relied on external operators or platforms to facilitate the shift. A major advantage of digitizing food vouchers was the ability to monitor

Table 3.3 Changes in food voucher formats between the first and second pandemic waves (99 cities, April 2020 to April 2021)

Number of municipalities changing voucher format between the first and second food voucher delivery
Total 99 municipalities

Voucher format	*First pandemic wave*	*Second pandemic wave*			
		Paper	*Digital*	*Digital + paper*	*Other*
Paper	70	28	38	2	2
Digital	27	3	22	1	1
Paper and digital	5	1	4	0	0
Other	5	2	2	0	1

Source: Made by the author based on the original database 'Alimentaria'

and adjust their use in real time at participating stores, thus ensuring compliance with program guidelines and verifying eligible purchases.

Table 3.3 provides a detailed breakdown of the net change in voucher formats from the first to the second distribution. The second column represents the absolute frequencies of voucher format used in the first round, while subsequent columns indicate how each format changed in the second round. For instance, the 'paper' row shows that 70 municipalities used only paper vouchers in the first round. By the second round, only 28 retained paper vouchers, two switched to other formats, two used both paper and digital vouchers, and a substantial 38 municipalities transitioned to digital vouchers. Only one-third of municipalities maintained home delivery during the second wave, compared to 63% in the first wave. The shift to digital vouchers constitutes a broader transformation in social service delivery, balancing large-scale digitization with adaptability to community-specific needs. Research has explored how digital strategies can both empower and disadvantage citizens (Zerrer & Sept, 2020; Eyieyien et al., 2024). While digital vouchers enhance efficiency, accessibility, and fraud prevention, they also present challenges for individuals with limited digital literacy or access to technology. This research does not specifically investigate this aspect. However, municipalities took different approaches to digitization: some invested in training information systems managers and frontline workers to implement new digital processes, while others relied on external operators, revealing gaps in organizational digital capacity. This suggests that digital transformation in public services is not just about adopting new technologies but also about ensuring that local institutions have the internal expertise to manage and support these systems. Municipalities that prioritized staff training were better positioned to handle technological disruptions, troubleshoot issues in real time, and adapt their digital services to changing needs. In contrast, those that outsourced digitization processes risked dependency on third-party providers, which could lead to higher costs, reduced flexibility, and potential service disruptions if contracts ended or external systems failed to integrate smoothly with municipal workflows. Overall, these insights highlight the need for proactive, long-term planning, investment in digital infrastructure, and adaptive learning mechanisms to increase the effectiveness and resilience of social and food services during crises.

Collaboration with External Organizations

According to the National FAP (Food Assistance Program) design, local street-level bureaucrats (SLBs) are expected to address information asymmetries by engaging both public and private actors who possess strategic or complementary knowledge about the target population. This collaboration was crucial in designing and implementing an effective take-up process. By fostering closer cooperation and communication among different agencies and actors, SLBs can significantly reduce the barriers that prevent eligible individuals from accessing the support they need. The evolution of food voucher distribution, based on collaborative networks, saw significant changes from the first to the second wave of the pandemic at the urban level. During the initial wave, there was a high level of collaboration between public and private street-level operators. Most of the workforce came from 'Civil Protection and local police officers,' supplemented by personnel from the third sector, voluntary organizations, the Red Cross, and voluntary associations. This collective effort was critical in responding swiftly to the crisis. However, by the second wave, the nature of these collaborations changed. Most personnel were appointed directly by the municipality or local police, and collaboration with the Red Cross came to an end. This change also led to a major reduction in the involvement of third-sector and voluntary-sector personnel. Despite national legislation that incentivized public–private collaboration and granted special status to third-sector entities and volunteers during the pandemic, the practical execution at the local level moved toward an operation that was more overseen at the municipal level. Interviews with organizations such as the Red Cross (RC), Food Banks, and Caritas reported an increasing number of 'irregular' and 'new vulnerable' individuals seeking assistance. These groups often included people who had yet to interact with social services, highlighting these networks' growing reach and importance. The Italian National Red Cross gained valuable insights during the lockdowns through its collaboration with municipal food voucher and food bag delivery programs. In response to the evolving needs, the Red Cross, in partnership with a major national distribution player (as noted in Interview 5), adopted a new tool (a booklet of food vouchers) for the first time to better assist those in need. This new approach would have given RC beneficiaries greater autonomy in choosing what to buy and eat, providing more dignified and flexible assistance. This initiative reflects a broader trend toward enhancing the effectiveness and responsiveness of food assistance programs within public–private partnerships, enabling them to adapt more efficiently to the evolving landscape during turbulent periods. During the first wave of the pandemic, municipal social services extensively leveraged external relationships and collaborations to manage the food voucher distribution and delivery system. In many cities, volunteer brigades were mobilized to ensure effective outreach to needy households. However, a significant shift occurred in the personnel responsible for food voucher distribution from the first to the second pandemic wave. Much of the support staff initially came from Civil Protection units and other categories, with substantial contributions from the third sector, volunteers, and the Red Cross. As the crisis progressed, the governance of urban food security transitioned from an open, multi-agency relational system—where external actors such as the Red Cross, volunteers, and third-sector organizations played a pivotal role—to a more traditional model. In this latter phase, the involvement of third-sector organizations and the Red Cross diminished, while the roles of municipal officials and local police were significantly enhanced. Trust-building between municipal street-level officials and external stakeholders proved critical for effective crisis management. The strong initial collaboration between public and private actors, including the Red Cross and third-sector organizations, enabled a swift and comprehensive response to the population's immediate needs. However, the decline of these partnerships during the second wave highlights challenges in institutionalizing long-term cooperation. Maintaining trust between municipal officials and external stakeholders is crucial for the resilience and effectiveness of social services. Strengthening and maintaining trust between municipal officials and external stakeholders is vital for social services' long-term resilience and effectiveness and one may argue that local governments with a long-standing practice of collaboration with the third sector are more likely to institutionalize increasing and diverse networks. The shift from broad, multi-agency efforts to more municipal-driven responses reflects an adaptive learning process within social services. This process involved assessing the strengths and limitations of expansive partnerships, ultimately leading to more centralized strategies. By documenting and analyzing these collaborative experiences, municipalities and social services can develop adaptable best practices and frameworks, thus ensuring they are better prepared to implement effective, coordinated responses in future crises.

3.5 Streamlining Administration During the Pandemic: Enduring Reforms or a Return to Routine?

An additional aspect worth considering is the intense effort by social services to simplify procedures for accessing funds, including extending the opening hours of physical and virtual branches. Digitization clashed with the limited spread of digital identities among end beneficiaries (digital divide) and the local governments' poor functioning of enabling platforms. Consequently, 'facilitators' had to be employed to resolve beneficiaries' problems on a case-by-case basis through telephone contact. Interestingly, some of the simplified access measures have transitioned into permanent services for vulnerable populations, particularly Ukrainian immigrants and refugees. These stable services now include food aid, housing assistance, and other types of support. Digitization is an opportunity to return to more routine, simplified processes that were proven effective during the pandemic, ensuring that the lessons learned continue to benefit those in need. Again, municipal social services and charitable organizations adopted simplification and digitization of the database for individuals experiencing food insecurity. A notable example is the Red Cross, which learned during the pandemic that integrating data across various charitable and third-sector agencies could benefit those in need and expedite responses. For the first time in the Veneto region, the Red Cross created a single database, integrating its data with that of Caritas and the municipal parishes, which coordinate activities with the social services of local governments. The digitization and integration of databases streamlined the processes for identifying and assisting needy individuals. By consolidating information from multiple sources, organizations were able to quickly cross-reference data and reduce redundancy, leading to faster and more accurate service delivery. This improved coordination among various entities, ensuring that resources were allocated more efficiently and effectively. The shift toward digitization required organizations and SLBs to adapt their routines and workflows. Social service operators and charitable organizations had to develop new skills. The integrated CR database's success in the Veneto region serves as a model for other regions and organizations. The sustainability of this digital approach ensures that it can be scaled up to include more agencies and services, further broadening its impact. This model exemplifies how digitization can improve food assistance and support systems for vulnerable populations. In summary, simplifying and digitizing data management for food assistance improved immediate responses during the pandemic and set the stage for more efficient and coordinated food assistance in the future.

The FAP has created a direct interaction between central resources and local governments, cutting out the regions. Therefore, many regional governments have adopted autonomous initiatives to combat food insecurity by making autonomous funds available (Cattivelli & Rusciano, 2020). The COVID-19 emergency has also pushed Italian regions to implement unprecedented measures to address food insecurity. These experiences have highlighted the importance of an integrated approach involving public funds, local initiatives, and collaboration with the third sector. The solutions adopted, such as food vouchers, emergency funds, and direct distribution of food, have helped reduce the impact of the economic crisis on millions of Italian families, demonstrating the ability of regional institutions to react promptly. In the Veneto region, the Network of Solidarity Stores project has been included for many years among the regional policies for combating poverty and promoting social inclusion. The project was created with a regional law in 2011 to promote the recovery of excess food and its redistribution to those who assist people in severe social hardship and poverty using Third Sector Entities. The activities of the Network have been consolidated and increasingly developed over the years, also thanks to the impetus given by Law no. 166/2016 (Gadda Law). The involvement of Federdistribuzione was also very important, with which, in 2017, the Veneto Region signed a Collaboration Protocol to promote actions in the regional territory to encourage and enhance the donation of excess food. In 2018, the project was included in the framework of the 2018–2020 Regional Plan for Combating Poverty, adopted with Decision of the Regional Council No. 1504/2018 (updated with Decision of the Regional Council No. 1193/2020)[5] continued with the 2021–2023 poverty reduction program. The 'Empori della Solidarietà' enables families in need to buy food through a points card in organized places, such as a retail store, thus introducing an alternative method to the distribution of food parcels. The Network has grown over the years: in 2015, only 7 Emporia were part of it; in 2019, 24; in 2020, the network had grown to 26 Emporia, and in 2021, the number of 28 consolidated entities was reached. Other regions, such as Lombardy, Campania, and Lazio, have made financial resources available to purchase food vouchers for needy households. Other regional initiatives were devoted to connecting local producers and consumers (Tuscany region).

3.6 From Food Aid to Food Income: A Shift in Approach

The shift from food aid (such as in-kind food distributions and vouchers) to food income (direct cash transfers or subsidies for food purchases) constitutes a significant transformation in welfare governance. In December 2022, nearly two years after the outbreak of the pandemic, the Italian government launched a pilot Food Income Initiative (FII) to support individuals living in absolute poverty. Facilitated through third-sector organizations, the initiative provides free food parcels made from unsold products donated by retailers. These parcels can be reserved via a mobile app and either collected at designated distribution centers or delivered directly to vulnerable members of the population. Spearheaded by the Minister of Labor and Social Policies under the Meloni Government, the initiative is part of a broader strategy to combat severe poverty through food distribution. It builds upon previous programs such as the PO I FEAD Program (2014–2020) and extends into the National Inclusion and Poverty Reduction Program (2021–2027). The FII will be piloted over three years in capital municipalities of metropolitan cities, selected based on poverty levels, resource availability, and fair allocation criteria set by the Unified Conference. Third-sector organizations will manage food distribution, using products that are unsold but can still be consumed, such as items with damaged packaging or nearing expiration. Beneficiaries will include individuals identified by Local Partner Organizations (OpT in Italian) under the FEAD program, as well as those referred by local social services and third-sector organizations. The selected capital municipalities will submit Food Income project proposals in response to a public notice from the General Directorate for the Fight Against Poverty and Social Planning at the Ministry of Labor and Social Policies. These projects will prioritize collaboration with third-sector entities—particularly those already engaged in the FEAD program—and commercial businesses. Public funds allocated to the Food Income Initiative will complement resources under Priority 3: Counteracting Material Deprivation within the National Plan for Inclusion and the Fight Against Poverty (2021–2027). Additionally, part of these funds may be used to develop an IT application aimed at improving accessibility and usability, including features for tracking donated products and managing home delivery logistics. A dedicated working group will oversee the initiative's implementation. This group will include representatives from various ministries, ANCI (National

Association of Italian Municipalities), National Partner Organizations (OpN in Italian) in the FEAD program, the Intermediate Body of the Managing Authority of Operational Program 1 of the FEAD, as well as representatives from FIPE, FederDistribuzione, and Confcommercio. The Director of the General Directorate will chair the working group for the Fight against Poverty and Social Planning at the Ministry of Labor and Social Policies or their delegate. During the three-year pilot phase, the initiative was launched in Genoa, Naples, Palermo, and Florence to address food poverty and reduce food waste from large-scale retail operations. The municipalities will identify food sector operators willing to donate unsold goods, promotional surplus, and near-expiration products to third-sector organizations to be distributed. Applications to participate must be submitted through a form available on the Municipality's civic network. Another public notice will be issued to identify the third-sector organization responsible for managing the transport, storage, packaging, and distribution of food parcels. Applicants must provide proof of significant experience in similar activities, with the possibility of appointing a 'leader' to coordinate efforts with the Municipality. The primary innovation of the initiative lies not so much in the substantive aspect—such as the availability of new financial resources to support those in need—but rather in the procedural aspect. The real breakthrough is the effort to institutionalize public-private collaboration within cities, particularly by testing the 'Milan model' in other urban settings. Building on the FAP experience, the national FII seeks to formalize key lessons learned during the pandemic—most notably, the critical role of public–private partnerships in enhancing collaborative learning in the design and implementation of food security programs. A major advancement is the formal integration of third-sector organizations into the food distribution process, ensuring their systematic and sustained involvement. By leveraging these organizations' networks and expertise, the initiative enables a faster, more efficient, and targeted response to food insecurity. Ultimately, the FII institutionalizes public–private collaboration within a structured framework, strengthening resilience in times of crisis. Moreover, the FII integrates digital tools to enhance resource tracking and management in urban food security policies. By streamlining these technologies, the program improves accessibility and efficiency for both intermediary organizations and beneficiaries, ensuring smoother resource distribution. The selection of capital municipalities is based on data-driven criteria, considering poverty levels and equitable resource allocation. This approach, shaped by pandemic experiences, ensures targeted support for the most vulnerable areas, reducing food access disparities and promoting equity. A dedicated working group for FII monitoring underscores the program's commitment to oversight, evaluation, and continuous improvement. This focus on accountability and adaptability, reinforced during the pandemic, ensures the initiative remains responsive to emerging challenges. The FII emphasizes community engagement as a key factor in addressing food security. The pandemic demonstrated that local organizations are best positioned to identify and meet their communities' unique needs. Consequently, the initiative promotes localized, community-driven solutions, empowering those closest to the issue to lead the response. This approach is especially urgent, as 2023 statistics reveal a rise in material and social deprivation in Italy. 4.9 million Italians (8.4% of those over 16) cannot afford a full meal every two days, underscoring the need for targeted, locally led interventions.

The FII marks a significant shift in Italy's approach to food security, moving from short-term food aid toward a more systemic income-based model. The COVID-19 pandemic forced the government to reassess its priorities, emphasizing food security and reallocating resources to protect vulnerable populations. While the crisis initially drove interagency collaboration and community partnerships, these efforts were often seen as temporary measures rather than long-term structural changes. Governance mechanisms also evolved during this period. Street-level bureaucrats rapidly adopted digital tools and engaged with local organizations to enhance service delivery. However, this exposed the need for sustainable institutional support and a more structured learning process within public administration to prepare for future crises.

It remains uncertain whether the 2022 Food Income Initiative will serve as a precursor to a comprehensive national strategy recognizing the right to food or whether it will remain a localized experiment dependent on sub-national initiatives with varying consistency. A fragmented approach risks limiting the program's long-term impact owing to inconsistencies in implementation, resource allocation, and institutional support. If the latter is the case, these initiatives may remain sporadic and experimental, potentially requiring more coordination and scale for a sustained, nationwide impact. For this initiative to evolve into a robust national framework, it must move beyond short-term, localized actions and ensure: (a) strong political commitment to recognize food security as a fundamental right; (b) cross-sectoral collaboration between public institutions, private actors, and civil society; (c) sufficient and sustained resources to scale initiatives effectively; (d) institutional mechanisms to standardize and monitor local implementations. The success of this transition hinges on whether food income can be institutionalized as a sustainable governance model, ensuring long-term food security beyond crisis-driven responses.

Notes

1. To assess the food needs of recipients during the pandemic, the Ministry of Labor held a virtual meeting with representatives from the National Coordination Structures of the Lead Partner Organizations. These included Associazione Banco Alimentare Roma, Banco delle Opere di Carità, Caritas Italiana, Comunità di S. Egidio, Croce Rossa Italiana, and Fondazione Banco Alimentare. The meeting aimed to determine the contents of the food 'basket' to be purchased and distributed to those in need. https://www.lavoro.gov.it/temi-e-priorita/europa-e-fondi-europei/focus-on/fondo-di-aiuti-europei-agli-indigenti-Fead/Documents/RAA-FEAD-2020.pdf
2. https://www.lavoro.gov.it/temi-e-priorita/europa-e-fondi-europei/focus-on/fondo-di-aiuti-europei-agli-indigenti-fead/pagine/default
3. Human Rights Watch, World Report (2023). https://www.hrw.org/world-report/2023/country-chapters/italy
4. Dipartimento Protezione Civile, *Ulteriori interventi urgenti di protezione civile in relazione all'emergenza relativa al rischio sanitario connesso all'insorgenza di patologie derivanti da agenti virali trasmissibili*, Ocdpc n. 658 del 29 marzo 2020 Department of Civil Protection, *Additional urgent civil protection interventions in relation to the emergency concerning the health risk related to the outbreak of diseases resulting from transmissible viral agents*, Ordinances of the head of the Department of Civil Protection No. 658 of March 29, 2020.
5. BUR (Official Regional Bulletin), Regional Council Resolution No. 1504 of October 16, 2018, Legislative Decree No. 147./2017, Art. 14. *Regional plan for the fight against poverty. Decisions.* (Updated with Regional Council Resolution No. 1193/2020).

References

Abramson, M., & Mittu, R. (2006). Multi-agent coordination in open environments. In *Coordination of large-scale multiagent systems* (pp. 217–229). Springer US.

ActionAid. (2020). La pandemia che affama l'Italia. Covid-19, povertà alimentare e diritto al cibo, Roma, ActionAid.

Action Aid. (2023). *Frammenti da ricomporre.* Action Aid, Milano.
Action Aid. (2024). *I numeri della povertà alimentare in Italia a partire dalle statistiche ufficiali.* Action Aid, Milano.
Ansell, C. (2012). Collaborative governance. In *The Oxford handbook of governance.* https://doi.org/10.1093/oxfordhb/9780199560530.013.0035.
Ansell, C., Doberstein, C., Henderson, H., Siddiki, S. and 't Hart, P. (2020), Understanding inclusion in collaborative governance: A mixed methods approach, Policy and Society, 39(4), 570–591, doi:https://doi.org/10.1080/14494035.2020.1785726.
Ansell, C., & Gash, A. (2008). Collaborative governance in theory and practice. *Journal of Public Administration Research and Theory, 18*(4), 543–571.
Ansell, C., Sørensen, E., & Torfing, J. (2021). The COVID-19 pandemic as a game changer for public administration and leadership? The need for robust governance responses to turbulent problems. *Public Management Review, 23*(7), 949–960.
Bandera, L., Lodi Rizzini, C., & Maino, F. (2016). La povertà alimentare. *il Mulino, 65*(2), 259–267.
Barzelay, M. (1992). *Breaking through bureaucracy: A new vision for managing in government.* Univ of California Press.
Barzelay, M., & Thompson, F. (2010). Back to the future: Making public administration a design science. *Public Administration Review, S295*, 70.
Busetti, S. (2019). A theory-based evaluation of food waste policy: Evidence from Italy. *Food Policy, 88*, 101749.
Busetti, S., & Righettini, M. S. (2023). Policy learning from crises: Lessons learned from the Italian food stamp programme. *Policy & Politics, 51*(1), 91–112.
Capano, G., Howlett, M., Jarvis, D. S. L., Ramesh, M., & Goyal, N. (2020). Mobilizing policy (in)capacity to fight COVID-19: Understanding variations in state responses. *Policy and Society, 39*(3), 285–308. https://doi.org/10.1080/14494035.2020.1787628
Capano, G., & Lippi, A. (2021). Decentralization, policy capacities, and varieties of first health response to the COVID-19 outbreak: Evidence from three regions in Italy. *Journal of European Public Policy, 28*(8), 1197–1218.
Cattivelli, V., & Rusciano, V. (2020). Social innovation and food provisioning during COVID-19: The case of urban–rural initiatives in the province of Naples. *Sustainability, 12*(11), 4444.
Currie, J. (2004). *The take up of social benefits.* NBER Working Paper Series 10488, National Bureau Of Economic Research, Cambridge, Massachussets.
Davidovitz, M., Cohen, N., & Gofen, A. (2021). Governmental response to crises and its implications for street-level implementation: Policy ambiguity, risk, and discretion during the COVID-19 pandemic. *Journal of Comparative Policy Analysis: Research and Practice, 23*(1), 120–130.

Elmore, R. F. (1979). Backward mapping: Implementation research and policy decisions. *Political Science Quarterly, 94*(4), 601–616.
Emerson, K., & Nabatchi, T. (2015a). *Collaborative governance regimes.* Georgetown University Press.
Emerson, K., & Nabatchi, T. (2015b). Evaluating the productivity of collaborative governance regimes: A performance matrix. *Public Performance & Management Review, 38*(4), 717–747.
Emerson, K., Nabatchi, T., & Balogh, S. (2012). An integrative framework for collaborative governance. *Journal of Public Administration Research and Theory, 22*(1), 1–29.
Eurocity. (2021). Food aid in European Cities. Working Group Food Policies, March, Brussels, 1–31.
Eyieyien, O. G., Idemudia, C., Paul, P. O., & Ijomah, T. I. (2024). The impact of ICT projects on community development and promoting social inclusion. *International Journal of Engineering Research and Development, 20*(7), 300–310.
Fiocco, G. (2004). *L'Italia prima del miracolo economico. L'inchiesta parlamentare sulla miseria (1951–1954).* Lacaita Editore.
Galanakis, C. M., Rizou, M., Aldawoud, T. M., Ucak, I., & Rowan, N. J. (2021). Innovations and technology disruptions in the food sector within the COVID-19 pandemic and post-lockdown era. *Trends in Food Science & Technology, 110*, 193–200.
Giest, S. (2025). The intersection of digital and social infrastructures in (a) spatial policymaking. *Policy Sciences, 58*, 1–16.
Gofen, A., & Lotta, G. (2021). Street-level bureaucrats at the forefront of pandemic response: A comparative perspective. *Journal of Comparative Policy Analysis: Research and Practice, 23*(1), 3–15. https://doi.org/10.1080/13876988.2020.1861421
Howlett, M. (2019). Temporality and the analysis of policy processes. In K. Goetz (Ed.), *The Oxford handbook of time and politics* (pp. 1–21). Oxford University Press.
Hupe, P., & Buffat, A. (2014). A public service gap: Capturing contexts in a comparative approach of street-level bureaucracy. *Public Management Review, 16*(4), 548–569.
ISTAT. (2021). *Poverty in Italy. Year 2021.* Report, National Institute of Statistic Press. Rome.
ISTAT. (2024). *Resta stabile la povertà assoluta, la spesa media cresce ma meno dell'inflazione.* National Institute of Statistic Press, Rome.
Kingdon, J. W., & Stano, E. (1984). *Agendas, alternatives, and public policies* (Vol. 45, pp. 165–169). Little, Brown.
Kleven, H. J., & Kopczuk, W. (2011). Transfer program complexity and the take-up of social benefits. *American Economic Journal: Economic Policy, 3*(1), 54–90.

Klijn, E. H., & Koppenjan, J. (2016). The impact of contract characteristics on the performance of public–private partnerships (PPPs). *Public Money & Management, 36*(6), 455–462.

Knill, C., & Tosun, J. (2020). *Public policy: A new introduction*. Bloomsbury Publishing.

Lodge, M. (2015). Accountability and consumer sovereignty. In *Accountability and regulatory governance* (pp. 235–264). Springer.

Lodge, M., & Wegrich, K. (Eds.). (2014). *The problem-solving capacity of the modern state: Governance challenges and administrative capacities*. Oxford University Press.

Lotta, G., Maynard-Moody, S., & Musheno, M. (2023). Practicing knowledge-agency in highly vulnerable communities: The frontlines of governance in delivering Brazilian human services. *Public Management Review, 27*, 1–21.

Malandrino, A., & Demichelis, E. (2020). Conflict in decision making and variation in public administration outcomes in Italy during the COVID-19 crisis. *European Policy Analysis, 6*(2), 138–146.

Righettini, M. S., & Sbalchiero, S. (2017). The regulatory state and its variants in communications sector: Paradigms, strategies and ideas of universality and inclusion in Europe. *Stato e mercato, 37*(2), 247–282.

Sager, F., Thomann, E., & Hupe, P. (2020). Accountability of public servants at the street level. In *The Palgrave handbook of the public servant* (pp. 1–17). Palgrave Macmillan.

Scerri, P., Vincent, R., & Mailler, R. (2006). Comparing three approaches to large-scale coordination. In *Coordination of large-scale multiagent systems* (pp. 53–71). Springer US.

Scott, I., & Gong, T. (2021). Coordinating government silos: Challenges and opportunities. *Global Public Policy and Governance, 1*(1), 20–38.

Sohn, H., & Kwon, N. (2021). Issues associated with the use of National Health Insurance contributions in determining public policy program beneficiaries. *International Review of Public Administration, 26*(4), 337–352.

Thomann, E. (2015). Is output performance all about the resources? A fuzzy-set qualitative comparative analysis of street-level bureaucrats in Switzerland. *Public Administration, 93*(1), 177–194.

Thomann, E., van Engen, N., & Tummers, L. (2018). The necessity of discretion: A behavioral evaluation of bottom-up implementation theory. *Journal of Public Administration Research and Theory, 28*(4), 583–601.

Van Oorschot, W. (1991). Non-take-up of social security benefits in Europe. *Journal of European Social Policy, 1*(1), 15–30.

Zerrer, N., & Sept, A. (2020). Smart villagers as actors of digital social innovation in rural areas. *Urban Planning, 5*(4), 78–88.

CHAPTER 4

The Pandemic as a Game Changer in Urban Food Security Policies

The evolution of European urban food security initiatives reflects a growing integration of food systems into urban policy and planning over time. With the Urban Agriculture Movement (1970s–1980s), cities such as Amsterdam, London, and Berlin adopted initiatives to reconnect urban residents with food production and provide fresh and accessible produce (Morgan, 2009; Stierand, 2012; Doernberg et al., 2019). Food security began to appear in urban policies during the 1990s (Sonnino, 2019). For example, initiatives in the United Kingdom focused on combating food poverty and integrating food into broader urban planning programs. In the 2000s, European cities increasingly adopted comprehensive food strategies to experiment with new governance modes that aim to integrate different sectors and actors (Vara-Sánchez et al., 2021; Moragues-Faus, 2021). Initiatives such as the Milan Urban Food Policy Pact (2015) formalized global city commitments to address food security, sustainability, and resilience (Filippini et al., 2019; Candel, 2020). The expansion of urban food security programs in European cities is also deeply embedded in the social dynamics that resulted from the 2008 financial crisis (Konstantinidis, 2024; Skordili, 2013; Zerbian & De Luis Romero, 2023).

It is widely recognized that the COVID-19 pandemic opened a unique window for European cities to rethink and transform how they tackle food insecurity, involve communities, and prioritize solutions for joint good and long-term sustainability (López Cifuentes & Fiala, 2022: 2). This

M. S. Righettini, *The Governance of Food Security in Europe*,
https://doi.org/10.1007/978-3-031-98083-1_4

shift placed a strong emphasis on involving a wide range of stakeholders in food system decision-making, making the process more inclusive and resilient to future challenges. However, establishing formal urban food governance (such as City food councils) does not directly lead to effective multi-agency collaboration. The analysis shows that the lack of clearly defined roles and expectations, monitoring tools, and procedures to address obstacles in operationalizing actions are crucial to avoiding adverse effects in coordinating urban food governance. As mentioned by food governance scholars, the lack of some mechanisms can frustrate stakeholders, undermining effective collaboration and implementation of actions (Mansfield & Mendes, 2013; Moragues-Faus & Morgan, 2015).

The pandemic highlighted the complexity of multi-agent collaboration within urban and rural food systems, underscoring the need to align food security policy goals, proposed solutions, and expected outcomes within local communities. A recent FAO report builds on and expands on the content of a previous report and shows that the pandemic has highlighted complex connections in food security, which require integrated and multilevel approaches to food governance (FAO, 2019, 2023). Collaboration includes public, private, and civil society sectors and stronger partnerships between local governments and national authorities. The FAO Urban Food Agenda is a key initiative to promote sustainable development, food security, and nutrition in urban and peri-urban regions and nearby rural areas. The commitment to food security ought to encompass various policies, programs, and initiatives developed and implemented through partnerships with multiple stakeholders, including civil society, academic institutions, UN and international agencies, city networks, and relevant public and private organizations. The focus on local governance and collaboration highlights a noteworthy trend: urban areas have become influential hubs of food system innovation, especially when they prioritize residents' well-being and environmental health.

As pointed out by Polman and Bazzan (2023), before the pandemic, the Milan Urban Food Policy Pact (MUFPP), signed in 2015 during the Universal Exposition dedicated to food by representatives from 132 cities, prompted inter-city cooperation in making urban food systems more sustainable, both environmentally and socially. Polman and Bazzan highlight that the Milan Urban Food Policy Pact (MUFPP) has sparked innovation in urban food governance. They identify five critical governance tools that are essential for crafting progressive urban food policies: multi-stakeholder collaboration, the integration of government efforts, the development of cohesive urban food strategies, and the monitoring of local initiatives. However, their analysis reveals a crucial insight: urban food governance innovation only depends on the presence of these tools. Instead, specific combinations of these tools can help to learn and, as a result, limit the potential for meaningful innovation (Polman & Bazzan, 2023). The finding underscores the idea that successful collaboration within FAP governance systems is not solely dependent on the presence of individual tools or strategies. Instead, it relies on the interplay and combination of tools to foster effective learning and innovation. Combining governance tools—such as multi-stakeholder collaboration, integrated efforts, and monitoring—creates a dynamic environment where learning can thrive. When such tools are well integrated, they may enable feedback loops, shared knowledge, and joint problem-solving, essential for continuous adaptation and improvement in policy and practice.

The Milan Urban Food Policy Pact (MUFPP) has significantly propelled dynamic processes that catalyze the growth of multi-stakeholder collaboration and networking in various European cities. In this context, Eurocities Food has emerged as an innovative city self-coordination mechanism, driving food policy advocacy and governance at the European level.

4.1 Urban Food Security Governance: A Cornerstone of Municipal Social Equity and Welfare

Since the early twenty-first century, evidence suggests that household food insecurity has risen significantly across many European countries (Loopstra et al., 2015; Davis & Geiger, 2017). In response, food aid—both direct and indirect—has become a key component of local welfare systems aimed at addressing food deprivation and vulnerability. This work does not engage in the broader debate on welfare systems using macro-level data (Davis & Geiger, 2017). Instead, it focuses on pre-pandemic research (2003–2011) examining how different welfare systems responded to food insecurity during and after crises. Findings indicate that while food insecurity increased across all welfare systems, Eastern European countries had the highest overall rates. However, the most significant post-crisis rise occurred in the Anglo-Saxon system, contrary to expectations. This surge has been linked to policy changes, particularly increased conditionality in the UK benefits system (Loopstra et al., 2015), highlighting the role of governance tools in shaping vulnerability. Stricter eligibility requirements and welfare conditionalities may exacerbate food insecurity by restricting access to essential support during economic downturns.

Food aid initiatives play a crucial role in supporting individuals and households facing financial hardship and health risks by providing essential assistance to supplement their income and food intake (Davis & Geiger, 2017). These initiatives encompass a range of community-based support systems, including food stamps and vouchers, prepared onsite meals, and various forms of offsite food distribution, such as 'soup runs' (Lambie-Mumford & Dowler, 2015). Previous studies, such as those by Van der Horst et al. (2014) in the Netherlands, Schoneville (2018), Selke (2017), and Garthwaite (2015), further highlight the significance of food aid programs in local welfare in mitigating food insecurity. Much of the existing literature analyzes food insecurity and its determinants from a household perspective, aiming to identify effective solutions. Urban food governance is conceptually framed as a platform for good governance and inclusive, democratic engagement among multiple stakeholders, fostering a food system built on solidarity and capacity building (Sonnino, 2019). However, despite the growing body of research, significant knowledge gaps remain regarding the evolving dynamics of urban food aid governance at the street level. In particular, access to food aid during crises, procedural tools and the interaction between charitable efforts, local welfare systems, and economic operators in shaping food aid initiatives have yet to be fully explored.

The COVID-19 pandemic underscored the critical role of urban food security governance, highlighting the urgent need for coordinated action across societal sectors, governance levels, and policy domains (Maino & De Tommaso, 2022). In response, cities have increasingly embraced more sustainable food governance models, empowering communities to address food insecurity in deeper and more lasting ways. Additionally, the war in Ukraine has significantly disrupted global and local food access, driving up inflation and reducing the availability of key commodities such as wheat, maize, and sunflower oil. The conflict has also had an impact on the production and trade of fuel and fertilizers, thus further exacerbating food insecurity (GRFC, 2023).

Amid these challenges, the growing focus on urban governance is closely linked to the need for validated, scalable methodologies to assess and address food insecurity in urban contexts (Crush & Fayne, 2010; Loh & Agyeman, 2019; Moragues-Faus & Battersby, 2021). There is a debate about whether data on income and food consumption are sufficient to capture the many social facets of the problem, such as the one relating to minors and adolescents (AA, 2022). With more than half of the global population and 75% of Europeans living in urban areas, and as much as 70% of the world's food production designated for consumption in urban environments, city authorities can use their administrative powers and proximity to citizens to tackle food insecurity, address inequalities, support healthy diets, and create shorter food supply chains (Zerbian & De Luis Romero, 2023). Many cities have strategically enhanced local food systems through dedicated policies and strategies, enabling them to respond promptly to the social consequences of the COVID-19 crisis. The crises underscored the value of integrated approaches to urban food systems. European funding was pivotal in advancing integrated strategies, fostering cooperation, and enabling the exchange of best practices among national and subnational governments. The collaborative effort has laid the groundwork for more resilient, adaptable urban food systems, demonstrating how effective governance can address complex challenges at the city level (Eurocities, 2021).

Funding streams like the European Fund for the Most Deprived (FEAD), the European Social Fund (ESF+), and the European Child Guarantee—focused on combating child poverty—have been pivotal in addressing urban food insecurity. Programs such as URBACT also addressed various food-related challenges, often requiring collaboration between different government departments or the creation of new governance entities. Urban food security policies frequently involve targeted initiatives with specific objectives, such as addressing public health issues like obesity or tackling environmental concerns like food waste (Busetti & Pace, 2023). These initiatives were important first steps toward broader, fully integrated food policies that could positively impact other areas, including agriculture, waste and water management, and education (Righettini & Lizzi, 2020).

While urban food security policies incorporate input from civil society and various stakeholders, they differ from grassroots, volunteer-led initiatives that operate independently of government efforts. Although these grassroots actions are valuable, they do not constitute formal public urban food policies. As emphasized by the EU, cities must adopt a more holistic and integrated approach, recognizing the complex socio-environmental and economic interconnections that are essential to develop effective crisis-response policies (Lang et al., 2009). Urban food security programs are just one component of broader efforts to transform urban food systems within a multilevel governance framework. Addressing food security requires a comprehensive approach that integrates public security and social equity considerations. Effective governance is essential in coordinating policies across multiple tiers—local, national, regional, and global—ensuring alignment between grassroots initiatives and institutional strategies. Strengthening governance mechanisms can enhance the resilience and sustainability of urban food systems, making them more responsive to crises and long-term challenges (Sonnino, 2023). Cities play a crucial role in addressing food security, leveraging their unique powers and street-level responsibilities. However, their efforts are shaped by decisions at higher levels of governance (Zerbian & De Luis Romero, 2023). Many food system challenges—such as trade policies, economic regulations, agricultural frameworks, and public health initiatives—are influenced by national and international imperatives. Despite these external influences, cities are increasingly at the forefront of tackling food insecurity. Urban governance serves as the primary arena where food-related challenges are addressed, allowing for more responsive, localized solutions. This dual responsibility underscores the complex interplay between local, national, and global policies in shaping urban food security. It also highlights the vital role of cities in pioneering innovative strategies of welfare to enhance food resilience and equity.

Ongoing commercial conflicts, the Russia-Ukraine war, and multiple regional crises are key drivers of acute food insecurity (Borràs-Pentinat, 2020). During periods of poly-crisis—whether triggered by war, pandemics, or natural disasters—policymakers must balance necessary restrictions with public security. While measures such as evacuations and lockdowns are essential for saving lives, they can also restrict personal freedoms and complicate responses to food insecurity (Cheek et al., 2022). This tension between public health measures and individual liberties highlights the weakness of urban food crisis management.

The restrictive measures implemented during the COVID-19 pandemic effectively slowed the virus's spread but also led to significant economic and psychological challenges. Urban food aid policies emerged as essential interventions to mitigate financial losses while addressing security risks. As lockdowns intensified, food aid management became a crucial countermeasure, with policymakers, businesses, and nonprofits working to ease the unintended consequences—particularly job losses and economic hardship (Crayne, 2020). As a result, food insecurity emerged as a critical urban governance challenge during the pandemic, exposing the limitations of welfare state protections. In many cases, minimum income support proved inadequate for vulnerable populations, revealing gaps in social safety nets and emphasizing the need for resilient, adaptable food aid systems (Goedemé et al., 2019).

As an urban policy actor observed (Interview 5), '*Almost all cities, during the pandemic, were compelled to reflect on food poverty and find ways to address deprivation and vulnerability in urban contexts during the crisis.*'

The evolving landscape of urban food aid reflects a complex interplay between charitable initiatives, local welfare systems, and market-driven economic imperatives (Lambie-Mumford & Silvasti, 2020). The rise of the 'new economy of charity' underscores how local welfare mechanisms are adapting and changing (Schoneville, 2018). However, its integration into broader welfare policies remains underexplored. This highlights the need to examine the growing interconnectedness between public and private social support systems in times of crisis and to reassess how food aid is conceptualized and managed in urban settings.

4.2 A New Knowledge Paradigm: Urban Food Security Governance in Crises

The pandemic forced many large and medium-sized cities across northern, southern, and eastern Europe to rethink and redesign their food aid policies (Lam et al., 2022). In northern Europe, many cities adopted innovative approaches to food security, using digital platforms and local networks to coordinate aid distribution efficiently. Similarly, in southern Europe, cities adapted their food aid strategies to address the unique challenges of the pandemic, relying more heavily on community-based organizations and local volunteers to reach vulnerable populations (González-Azcárate et al., 2023; Jablonski et al., 2021). In eastern Europe, particularly in cities near the Ukrainian border, the war has severely disrupted traditional short food supply chains while simultaneously increasing the demand for provisions owing to the high influx of refugees (Kovács et al., 2022). This evolving landscape underscores the need for a more comprehensive approach to urban food governance, one that integrates multiple forms of knowledge: policy knowledge (institutional learning and adaptation), operational knowledge (best practices and program implementation), and social knowledge (community-led responses and resilience). Understanding how cities conceptualize and manage food aid within this broader governance framework is crucial for recognizing how different types of learning emerge and interact during crises. However, the effectiveness of these responses also depends on a city's ability—both in terms of institutional resources and local infrastructure—to adapt, coordinate, and support food aid initiatives. An interesting survey conducted in 2021 by Eurocities Food on managing food aid policies during the pandemic, involving 76 city officers from 47 European cities across 14 EU member states, identified varying 'levels of capacity' developed during the pandemic within urban food governance. The survey results allow for a better understanding of the mechanisms for urban food security programs' adaptation to the pandemic. The survey highlights a range of preparedness and expertise levels among urban food security street-level operators, reflecting different institutionalization and capacity-building stages within each country's urban food governance framework. The capacity ranges from well-established governance modes in countries such as France, Italy, and Spain to emerging or nascent capabilities in Germany, Greece, and Poland (Eurocities, 2021, p. 4). The differences reveal the varying ability of municipal street-level bureaucrats (SLBs) to monitor local food needs and coordinate effective responses. In cities that had well-established food governance systems before the pandemic, SLBs engaged in timely and frequent interactions with key stakeholders, showed in-depth knowledge of local food aid distribution channels, and more advanced experience in managing tools such as food vouchers. The survey shows that some cities were better equipped to adapt and respond swiftly to the pandemic regarding the number of public operators involved in food aid during the pandemic (Eurocities, 2021, p. 6). Cities with more human resources—public operators—engaged in food aid appeared to provide vital institutional support and a more robust interagency coordination structure in food aid programs, such as the ability to scale up food aid initiatives quickly and provide more comprehensive responses to food insecurity in times of crisis. Conversely, cities with fewer public operators involved in these activities relied more heavily on ad hoc or fragmented responses, meaning that there may be a need for further institutional capacity building. The survey further reveals that cities with food security responsibilities embedded within welfare departments or dedicated food policy offices proved to be more committed to food aid. This specialization reflects a governance mechanism in which food security is institutionalized within departments focused on social welfare, ensuring that food aid aligns with broader welfare goals. Such departmental anchoring enabled SLBs to leverage specialized knowledge and funding and established networks to address food insecurity more effectively. By embedding food security within welfare-focused units, cities could benefit from a more holistic, well-supported, and well-resourced approach to food aid that extends beyond emergency responses. The findings also underscored the importance of multilevel governance and cross-sectoral collaboration in urban food security. Cities with higher levels of knowledge and coordination between local actors, such as charity organizations, FEAD partners, and regional hubs, were better equipped to address food insecurity during the pandemic. Effective food security governance relies on a network of coordinated actors and multilevel partnerships, which enable cities to leverage resources and expertise across sectors. In contexts where these connections are less established or being developed, such as in Poland or Greece, the pandemic underscored the need to strengthen cross-sectoral and cross-level relationships to build more resilient food systems. The Eurocities Food report offers valuable insights into the variations in urban governance configurations and dynamics, highlighting significant disparities in knowledge, capacity, and preparedness among urban operators. These differences are exemplified by the contrast between France's well-developed and coordinated food security networks and Germany's comparatively fragmented approach (Eurocities, 2021).

France is a prime example of high operational capacity and institutionalized knowledge among street-level bureaucrats (SLBs) and city officials, as noted in the Eurocities survey results (Bouchet & Duvoux, 2023). Over the years, many French cities have built a robust framework for urban food security, blending national and EU programs with collaborations between public entities, local charities, and food aid organizations, such as Food Banks and Restos du Cœur. This framework provided a sound foundation during the pandemic, enabling cities to rapidly mobilize a network of diverse actors—including local producers, retailers, charitable organizations, and municipal centers for social action (C.C.A.S.)—to address sudden surges in food insecurity. Established platforms such as the *Territorial Food Project (PAT)* and digital tools for connecting local producers with consumers became instrumental in maintaining food access amid supply chain disruptions. Moreover, the crisis spurred innovative partnerships between large retailers and charitable organizations, which are typically uncommon in non-crisis contexts. These deep-rooted networks and collaborative mechanisms enabled French cities to implement multifaceted responses swiftly, exemplifying how established governance structures facilitate effective crisis management at the urban level.

In contrast, Germany's urban food security governance aligns with the survey's findings, highlighting its ongoing efforts to develop and enhance capacities. German city government officials have a general knowledge of local actors but have yet to gain experience with the FEAD food aid program (OP1) and its resources, such as hubs and public–private partnerships. They possessed an extensive understanding of local actors, however, they lacked direct experience with the FEAD food aid program (OP1) and its associated resources, such as hubs and public-private partnerships. This knowledge gap has led to a less institutionalized approach to food security governance, where mutual awareness among stakeholders rarely translates into coordinated actions or effective crisis responses. Although local government officials were familiar with individual FEAD projects, the absence of an integrated framework and collaborative networks hindered these initiatives from developing into a unified urban food security strategy, leaving cities needing more preparation to address systemic challenges. As a result, cities in Germany struggled to mobilize the same breadth of resources and interagency support that French cities could access during the pandemic, highlighting the importance of established governance mechanisms in enhancing urban resilience to food insecurity. France's longstanding collaboration with food aid organizations (Food Banks and Restos du Cœur) and structured food governance networks enabled its cities to leverage expertise, resources, and established relationships during the pandemic, addressing food insecurity swiftly and effectively. In Germany, the absence of a similarly institutionalized framework at the urban level limited cities' abilities to coordinate and upscale food aid responses, thereby underscoring the importance of building institutional capacity. In France, during the pandemic waves, collaboration networks were quickly activated, such as partnerships between C.C.A.S., charitable organizations, local producers, and large retailers. As is reflected in the Eurocities survey, Germany's limited cross-sector collaboration highlighted the need for closer partnerships and network building among stakeholders to achieve more resilient food security systems. In France, integrating programs such as the Territorial Food Project (PAT) has fostered system-wide learning, enabling cities to develop unified and adaptive responses to food security matters (HLPE, 2020). In Germany, reliance on isolated projects has limited the development of system-wide learning, which is essential for creating cohesive and scalable responses to food crises. The pandemic exposed vulnerabilities even within France's robust framework, such as shortages of volunteers and food supplies. However, the established urban governance mechanisms enabled French cities to adapt by enlisting a broader range of stakeholders, such as private-sector actors and local producers. In contrast, German cities, with less developed frameworks, were less equipped to adapt or expand their food aid operations when demand surged unexpectedly.

The Eurocities survey and insights into the French and German context reveal that a city's ability to effectively manage food security during crises may depend on preexisting public–private alliances, institutional knowledge, and inter-organizational networks.

This work's focus now turns to the governance dynamics of two European cities in the context of food security challenges that were exacerbated by crises: Milan, celebrated as Italy's economic powerhouse and a trailblazer in food policy innovation, and Warsaw, Poland's rapidly developing capital, where social and economic transformations intersect with rising vulnerabilities. These cities operate within distinct welfare systems that shape their approaches to food aid and social support. Italy follows a more decentralized, family-oriented welfare model, where municipalities and civil society play a crucial role in addressing social needs. In contrast, Poland's welfare system has historically been more centralized, with a stronger reliance on state-led interventions, though recent socio-economic shifts have prompted greater involvement of NGOs and local initiatives. Examining how these differing welfare structures influence urban food governance provides valuable insights into the adaptability and resilience of local food aid strategies. By comparing these cities, this study aims to uncover how their distinct contexts shape policy, operational and social learning, highlighting the impact of crises on their ability to adapt and respond effectively. The case studies share certain similarities but differ in some key aspects. Both cities are EU Member States participating in FEAD OP1, primarily concerning food aid. However, their governance contexts vary significantly in terms of welfare decentralization and food policy maturity: Milan has had a city food policy since 2015, whereas Warsaw lacks one. Comparing Milan and Warsaw means examining how different welfare systems and urban food policy frameworks respond to food insecurity during crises. This work explores whether anti-crisis programs and practices were layered onto existing systems, creating a hybrid governance model where old and new approaches coexist, or whether new solutions are integrated with existing frameworks to enhance overall governance capacity. This analysis combines secondary qualitative data—such as institutional documents, official reports, and statistics—with in-depth interviews with municipal operators to investigate urban food security governance at the street level. The case of Warsaw is particularly complex, as it spans two overlapping crises: the COVID-19 pandemic (2020–2021) and the humanitarian crisis triggered by the Russia-Ukraine conflict (2022–2023). For Milan, this work examines the city's response to the pandemic from 2020 to 2023.

4.3 Poland's Challenges in a Poly-crisis Era

Poland is still considered a centralized administrative system, but the reconstruction of local government was a key pillar of the 1989 political transformation. The country's 2004 accession to the European Union further reinforced the principles of self-governance, encouraging decentralization and expanding the role of local authorities in social policy implementation (Kulesza & SzeŚciŁo, 2012). Despite these shifts, Poland's governance structure remains more centralized compared to countries such as Italy, where municipalities have greater autonomy in shaping social and food policies. Understanding these historical and institutional differences is crucial for analyzing how cities such as Warsaw and Milan develop and implement food aid strategies within their respective governance frameworks.

Russia's war against Ukraine began on February 20th, 2014, following the 'Revolution of Dignity' and the ousting of pro-Russian President Viktor Yanukovych. On that day, Russian troops illegally crossed Ukraine's border, leading to the occupation of Crimea. The conflict continued in eastern Ukraine and the partial occupation of the Donbas for eight years, escalating into a full-scale invasion in February 2022. This war, compounded by the COVID-19 pandemic, strained Ukraine's healthcare system and economy. The simultaneous crises exacerbated food insecurity and humanitarian challenges, especially for displaced populations and vulnerable communities. This interplay between war and pandemic underscored vulnerabilities in global supply chains and highlighted the critical need for adaptive governance in managing health and security emergencies. Poland, as an EU Member State, presents a unique case, facing consecutive crises that have tested its food security policies. The COVID-19 pandemic disrupted urban economic and social systems, reversing a declining unemployment trend as the rate rose from 5.2% in January 2020 to

6.2% in February 2021. This figure only partially reflects the impact, as many discouraged workers stopped looking for a job, and short-term unemployment often led to long-term inactivity. The pandemic also strained health and social systems, requiring immediate measures such as lockdowns, healthcare reorganization, and food security support for vulnerable populations.

Between 2014 and 2018, Poland made significant progress in reducing poverty and social exclusion, mainly owing to effective centralized social food aid programs. Under FEAD, organizations such as Food Banks and Caritas distributed over 252 million tons of food, alleviating hardship for vulnerable populations. In 2018, the program expanded to underserved areas, or 'white spots,' through greater involvement of local partners, such as social assistance centers, enhancing local coverage and impact. This groundwork proved essential in building resilience for future crises. The FEAD assistance in Poland effectively supported diverse, vulnerable groups, including children under 15, individuals with disabilities, the elderly, the homeless, those facing housing exclusion, economically disadvantaged individuals, low-intensity work households, the working poor, and residents of remote or underdeveloped areas. Local-level projects, particularly in cities, successfully targeted these groups, primarily focusing on children and individuals with disabilities needing food aid. Social workers and third-sector organizations played a vital role in implementing these initiatives. The program's success was underpinned by a multilevel governance approach involving urban and regional governments alongside other levels of administration. This coordinated strategy enabled comprehensive efforts to combat poverty and social exclusion, solidifying FEAD's role as a cornerstone of Poland's social welfare system.

Additionally, the passage points out the role of Poland's strong agricultural sector in the country's overall economic response during the crisis. With agriculture, forestry, and fisheries making up 2.6% of the Gross Value Added (GVA) in 2019 and a high percentage of farms being family-owned, this agricultural strength provided a critical foundation for the country's food assistance strategies during the pandemic. The passage underscores how social welfare initiatives and Poland's agricultural context worked together to address food insecurity and poverty during a challenging time.

The COVID-19 pandemic severely disrupted local food supply chains in Poland, causing shortages, price hikes, and logistical challenges that affected food availability and affordability, particularly for low-income households. Job losses and reduced incomes exacerbated food insecurity, disproportionately impacting vulnerable groups such as low-income households, the elderly, and marginalized communities. The primary support mechanism for Poland's economy, including the agri-food sector, was the 'Anti-Crisis Shield,' a comprehensive program offering targeted measures for food supply chain entities. These included social insurance premium exemptions, employment subsidies, price controls, tax breaks, loss deductions for pandemic-affected taxpayers, extended permits for foreign workers, simplified e-administration access, deferred loan payments, favorable credit terms, and enhanced work-time flexibility for agri-food entrepreneurs.

The pandemic also had severe social repercussions. The economic shutdown led to major job losses, with approximately 660,000 people losing their jobs within the pandemic's first three months. This surge in unemployment heightened financial insecurity, pushing some individuals into poverty and limiting their access to adequate nutrition. Mobility restrictions further exacerbated food access challenges, disproportionately affecting vulnerable groups such as the elderly, low-income households, and residents of isolated or under-resourced areas.

The Polish central government addressed severe poverty through the Food Aid Operational Program, providing various forms of food assistance. These initiatives responded to immediate crises and aimed at poverty alleviation, rural development, market competitiveness, and modernization across agriculture and food production sectors. During the COVID-19 pandemic, food insecurity increased, particularly among those struggling financially. In response, both large and small food producers launched projects to support vulnerable groups, such as the elderly and healthcare workers. One notable initiative, the 'Wzywamy Posiłki' (Meals Initiative), distributed over 220,000 meals to hospital employees and provided food aid to more than 10,000 senior citizens during the first six months of the pandemic. This collective effort reflected the broader commitment to addressing food insecurity during an unprecedented crisis.

The pandemic prompted significant shifts in consumer and food producer behavior patterns in Poland. In the first half of 2020, the country saw a rise in new online shops, with more farmers turning to online sales to connect directly with consumers. However, challenges such as limited internet access and quality concerns remained for some individuals. Platforms such as the 'Polish e-Bazaar' and initiatives such as KARMNIK emerged to bridge the gap, linking urban consumers with local producers and promoting the distribution of high-quality local food.

The pandemic sparked innovative responses from both producers and consumers in Poland. Farmers turned to social media platforms such as Facebook to share updates and build trust, fostering direct producer-consumer relationships. The cooperative Spółdzielnia Ostoja introduced 'vegemats,' parcel lockers for contact-free storage and delivery of organic vegetables. Additionally, the 'My Zbieramy' web portal enabled consumers to pick their own produce, reducing food waste, saving costs, and enabling farmers to harvest some of their crops. Some farmers also offered overnight stays and on-farm experiences, strengthening personal connections with consumers. In the aftermath of the pandemic, rising unemployment and economic instability worsened food insecurity among vulnerable populations.

The consecutive crises in Poland have highlighted its evolving approach to food security governance, positioning it as a key case study for understanding adaptive policymaking in complex, rapidly changing contexts. The overlapping challenges prompted swift action, including the creation and expansion of emergency food assistance programs in partnership with international organizations and NGOs. The coordinated response involved policy adjustments such as expanding distribution points, revising eligibility criteria, and ensuring safe food access under health protocols. Poland's rapid adaptation provides valuable insights into crisis management and learning processes in food security.

In 2022, Poland emerged as a leading country in providing support to Ukrainian war refugees despite the considerable strain on its national resources.

A Closer Look at Poly-crisis in Warsaw

Between 2020 and 2023, Warsaw faced a series of critical challenges. They began with two pandemic lockdowns, the first in March 2020 and the second in September 2020. Shortly thereafter, the city confronted a humanitarian crisis as it grappled with a significant increase in the influx of Ukrainian refugees. According to UNHCR data, the Russian invasion of Ukraine in 2022 displaced approximately 6.3 million refugees, with 5.9 million settling in Europe. Of these, 1.64 million applied for asylum in Poland, and around 800,000 were permanently registered in Warsaw. In Warsaw, the army carried out the initial emergency response, including home food deliveries and waste collection, in collaboration with municipal authorities. Public institutions adopted strict public health protocols, while civil society organizations provided critical support in food, healthcare, and mental health services.

During the humanitarian crisis, upon their arrival at Warsaw's train stations, Ukraine refugees were met with immediate assistance, including accommodation, food, transportation, and essential information to help them navigate their stay in Poland. The humanitarian crisis amplified the demand for food assistance and social services, putting additional pressure on cities that had already stretched their systems. Post-pandemic food security challenges and the influx of war refugees were among the most pressing issues for Warsaw's municipality and NGOs. These complex problems demanded immediate and sustainable solutions to support both the local population and Ukrainian refugees.

In Warsaw, the initial response to the refugee crisis saw humanitarian organizations and volunteers taking the lead in organizing aid at primary information desks. The city administration soon coordinated these efforts, launching platforms such as 'ochotnicy' and a dedicated donation mailbox to streamline support. Information desks staffed by multilingual personnel were set up, and funding was directed to the Multicultural Center, where refugees could access food, accommodation, legal counsel, employment support, and educational enrollment for children.[1] By the end of May 2022, over 338,000 people had been assisted with the help of more than 14,000 volunteers.

The Special Act enacted on March 12th, 2022 (retroactive to February 24th, 2022, and amended in January 2023) outlined the framework for refugee aid, including residency rights and access to essential services. Key support areas included financial aid (32%), food assistance (22%), accommodation (17%), WASH items (12%), clothing (7%), and transportation (5%). Poland's assistance to Ukrainian war refugees has unfolded in two phases: reception and integration. During the reception phase, refugees were given temporary assistance to address basic needs, including food aid through the Fund for European Aid to the Most Deprived (FEAD). This included packages or meals to ensure sustenance while refugees worked toward autonomy. Local governments focused on immediate humanitarian aid such as temporary housing, food provision, transportation, and guidance to navigate the new environment. The second phase shifted the focus toward the medium- and long-term integration of refugees into various facets of urban life. This involved efforts to integrate them into the labor market, the education system, and social welfare programs, as well as ensuring equal access to urban services such as housing. The goal during this phase was to move beyond mere survival and foster a sense of belonging and stability within the host communities.

The temporary nature of refugee aid under the Special Act primarily focused support on the reception phase, addressing immediate needs but often leaving full integration into society incomplete. Local governments had to adapt continuously, balancing short-term assistance with long-term integration efforts to meet refugees' evolving needs in a dynamic context. In Poland, special legislation has bolstered support networks and allocated resources to address the combined challenges of the COVID-19 pandemic and the war refugee crisis. This framework integrates food assistance into the broader social assistance system, prioritizing the urgent humanitarian needs of refugees. Municipalities are pivotal in implementing national directives, managing local responses, and ensuring effective food aid delivery. Poland's social assistance system, governed by the Law of March 12th, 2004, on Social Assistance, operates through multilevel collaboration involving central and local governments, NGOs, religious institutions, employers, and volunteers. Key components include municipal and commune social assistance centers (MOPS), private family support centers, and regional social policy centers in voivodeships, ensuring a coordinated approach to social and humanitarian aid.

At the city level, the Social Assistance Centers (MOPS) and Poviat Family Support Centers played a central role in delivering aid and offering financial support, such as cash benefits and non-financial assistance, including essential services. These centers addressed the needs of vulnerable groups, including those facing poverty, homelessness, domestic violence, parental loss, or natural disasters. NGOs significantly complemented this system by providing specialized services, including food aid, shelters for the homeless, homes for single mothers, day-care centers, and other support facilities. Collaboration between public institutions and nonprofits ensured a comprehensive range of services for those in need. Eligibility for social assistance benefits was primarily income-based, with municipal councils empowered to adjust benefit amounts for temporary or specific needs. Applications were processed through local MOPS, where social workers conducted a detailed assessment of personal, family-related, and financial circumstances within 14 days, ensuring targeted and efficient support.

The Polish central social assistance system provided many benefits, including financial aid, health and welfare insurance contributions, social work services, shelter, food assistance, care services, educational support, counseling, and family therapy. The system also included initiatives such as the Food Fund (Fundusz Alimentacyjny) for food assistance and programs such as the Karta Dużej Rodziny, which supported large families and agricultural businesses through food vouchers.

Two key departments within the Ministry of Family and Social Policy managed the benefits: the Assistance and Social Integration Department and the Department of Social Insurance. These entities ensured adequate resources and support to meet the population's diverse needs, particularly during crises. Poland's comprehensive approach, underpinned by a strong legal framework and multilevel governance, was instrumental in addressing the challenges of the COVID-19 pandemic and the war refugee crisis. Integrating food assistance into the broader social welfare system was proof of the country's commitment to meeting immediate humanitarian needs while fostering long-term social stability.

4.4 The Mechanisms of Food Assistance in Warsaw

Institutional Adaptability and Combined Response

The existing literature and empirical research on Polish policy-making and social policies remain limited. Current analyses emphasize the intricate dynamics between top-down transfers of social policy objectives and bottom-up initiatives. These dynamics underscore the growing importance of cities as key actors in multilevel governance, especially in the context of integration policies (Matusz & Pawlak, 2020; Wach & Pachocka, 2022). The COVID-19 pandemic and the Ukraine refugee crisis posed significant challenges to Polish municipal institutions and integration policies, prompting widespread procedural adaptations.

The pandemic had a profound impact on health and social systems, requiring immediate responses such as lockdowns, healthcare system reorganization, and food security measures for vulnerable populations. In Warsaw, the army led the initial national emergency response, including home food deliveries and waste collection, in collaboration with municipal authorities. Public organizations implemented strict health protocols, while civil society groups provided essential support in food, healthcare, and mental health services. In contrast, the Ukraine refugee crisis prompted a large-scale public–private response, with humanitarian organizations addressing the urgent needs of displaced individuals fleeing conflict. Polish institutions swiftly mobilized resources to provide shelter, food, transportation, healthcare, and other essential services. Temporary accommodations were set up, communication systems for refugees were organized, and coordinated actions involving government agencies, civil society, and international partners addressed immediate needs. Despite the different challenges, both crises highlighted the importance of institutional adaptability, inter-agency collaboration, and effective governance dynamics. The pandemic required a coordinated top-down response to control the virus's spread and support the resident population under lockdown, emphasizing centralized decision-making and top-down enforcement of public health measures. In contrast, the refugee crisis demanded urgent humanitarian aid for those seeking safety and shelter, fostering a more decentralized and collaborative bottom-up approach involving public, private, and civil society actors. Together, these responses not only demonstrated the capacity for effective collaboration across multiple sectors but also reinforced Poland's institutional resilience. The lessons learned from both crises are vital for enhancing future governance frameworks, ensuring that public institutions are equipped to handle a wide range of emergencies with adaptability, coordination, and a collective focus on urgent needs.

Unprecedented Volunteer and Social Organization Mobilization

During the pandemic crisis, individuals and organizations mobilized to volunteer across various roles, including preparing meals, distributing donations, managing sanitation, organizing children's activities, and supporting adults. Many volunteers were affiliated with established groups such as the Scout Movement (ZHP, ZHR), Caritas, or employee volunteering programs. In contrast, others signed up independently through email, online platforms, or social media. In larger cities, local governments facilitated volunteer registration; for instance, over 12,000 people joined the *Warsaw Volunteers* service. Despite the overwhelming response, coordinating such diverse and inexperienced groups posed challenges. Managing large numbers of volunteers with differing skills, schedules, and familiarity often proved inefficient. Pre-organized groups, such as high school clubs, were easily integrated—they were well-structured, arrived with mentors, and demonstrated strong commitment. The involvement of international organizations brought about both benefits and challenges. While their expertise was intended to improve operations, sudden leadership shifts sometimes created confusion. Decisions made by individuals unfamiliar with local systems led to inefficiencies, and the valuable experience gained by local teams during the pandemic remained underutilized.

Amidst multiple overlapping crises faced by Polish cities—particularly Warsaw—the collaboration between municipalities, volunteers, and non-governmental organizations (NGOs) proved essential for fostering effective and coordinated responses, highlighting key tools that enabled cooperation between institutional and non-institutional actors during such emergencies.

Creation of the 'Warszawa Wspiera' Support System: In response to the COVID-19 pandemic, the mayor of Warsaw launched the 'Warszawa Wspiera' Support System in March 2020. This initiative was designed to aid individuals in voluntary isolation or those in urgent need of support, with volunteers playing a crucial role. Collaborating closely with the municipality, these volunteers offered direct aid across Warsaw's districts, addressing essential needs such as grocery shopping, waste disposal, and pet care.

District-specific support teams[2]: Under the 'Warszawa Wspiera' initiative, district-specific support teams were established in early April 2020. These teams focused on assisting individuals in voluntary isolation, particularly the elderly and those at high risk of severe COVID-19 complications. The system benefited nearly 600 individuals, providing tailored assistance to meet their unique needs. This collaboration between volunteers, the municipality, and NGOs ensured the smooth operation of these district teams. The 'Warszawa Wspiera' teams included municipal staff and ov 500 volunteers from partnering non-governmental organizations. Residents could report cases of individuals needing help through the Municipal Volunteering Portal (www.ochotnicy.waw.pl) under the 'Residents' Support' section.

Collaboration with NGOs: The Warsaw municipality worked closely with various NGOs, including Caritas, Food Banks, faith-based organizations, and Soup Kitchen initiatives, to tackle food assistance challenges during the pandemic and the Ukrainian refugee crisis. Initially, 10 organizations were involved in food assistance efforts, but this number was later streamlined to five, ensuring a more focused and effective response. These NGOs received municipal funding to support their efforts in providing food aid to vulnerable populations.

Partnerships with private entities: The 'Warszawa Wspiera' initiative also established partnerships with private entities, including the Legia Foundation, the Girl Scouts and Scouts Emergency Service, the Creative in Action Foundation, and district Social Welfare Centers. These collaborations were pivotal in addressing the challenges of food assistance during the pandemic and the influx of Ukrainian refugees. Caritas, one of the most significant FEAD partners in food assistance, developed and implemented a food card system for elderly individuals in collaboration with the Jerónimo Martins Foundation. During the pandemic, Caritas operated numerous cooking and meal distribution centers across Poland, including 80 canteens, and provided food assistance through care packages. For the first time, a private foundation provided an annual grant of $5 million to Caritas for food assistance. Volunteers were essential in identifying and assisting potential beneficiaries with the application process, while electronic payment systems allowed for the monitoring of purchases to ensure compliance with guidelines.

Role of Food Banks: Poland has 31 food banks, slightly reduced from 32 owing to management challenges during the pandemic. These food banks collaborated with local governments and over 3000 NGOs nationwide to distribute food. The food supply chain was supported by 100 enterprises, including farmers, retailers, supermarkets, and producers. During the pandemic, the number of beneficiaries from organizations supported by Food Banks doubled, and the banks struggled to meet the surging demand. As a result, producers, faced with difficulties selling their products, became more active donors. Food banks also formed partnerships with foreign food companies to meet the increasing need for food assistance. These partnerships proved crucial in scaling up efforts during the pandemic and humanitarian crises, such as the Ukrainian refugee influx.

Enhanced collaboration within and across municipal administrative units: The crisis response efforts in Warsaw saw significantly strengthened collaboration among districts, driven by the city's unique administrative structure divided into cities and districts. This increased collaboration extended within districts and municipal offices, fostering improved coordination between key departments such as Social Security and the Security Office. Additionally, the central government of the capital and local offices worked more closely together, reflecting a collective effort to gather data and update social policies better to address the population's diverse needs during crises.

Lessons learned and future preparedness: After navigating three distinct crises, the municipality and NGOs have developed a strong sense of resilience and confidence in their ability to tackle future challenges, regardless of scale (Interviews 2—Municipal officer, and 3—Food Bank). According to those interviewed, the experiences gained from these crises have strengthened their capacity to manage food vulnerability more effectively in future emergencies. However, a key challenge in crisis management remains the sustainable utilization of volunteers. While short-term volunteer responses are often robust, maintaining engagement over the medium to long term has proven difficult. One municipal officer emphasized that while initial waves of volunteers are enthusiastic and eager to help, maintaining their involvement requires structured coordination and long-term incentives. A representative from the Food Bank highlighted that burnout is a major concern, as many volunteers face emotional and physical exhaustion over time. Additionally, volunteers may not always be deployed efficiently, leading to underutilization of skills or gaps in critical areas of aid distribution. These insights underscore the need for improved volunteer management strategies, such as structured training programs, better resource allocation, and partnerships that ensure a steady flow of support without over-reliance on short-term goodwill.

The key aspects of the relationships between public, private, and volunteer sectors can be categorized into three main areas: *institutional, operational, and social learning*. *Institutional learning* stems from effective coordination and clear communication among all stakeholders, which were essential to maximize the impact of volunteer efforts and avoiding duplication. The crises reinforced the need for structured collaboration between government agencies, NGOs, and private-sector partners to ensure efficient resource allocation. A systematic approach to learning from past experiences proved crucial. Stakeholders recognized the importance of institutionalizing crisis-response knowledge to enhance future preparedness and resilience. This reflection led to more adaptable volunteer management practices and better integration of volunteers into emergency planning. *Operational Learning* provides volunteers with adequate training and resources improved their ability to meet the demands of their roles, ultimately enhancing the quality and efficiency of crisis responses. The pandemic and refugee crisis exposed gaps in training, underscoring the need for structured capacity-building programs. Flexible and sustainable volunteer management approaches became a priority. The crises highlighted inconsistencies in coordination, insufficient training, and the lack of emotional and logistical support for volunteers in high-pressure situations. Moving forward, stakeholders prioritized capacity building, improved logistical frameworks, and stronger partnerships between volunteers, community organizations, and institutions. *Social Learning* regards maintaining volunteer motivation and commitment over time proved to be one of the greatest challenges, particularly as crises persisted. Strategies to maintain engagement included recognizing contributions, offering meaningful roles, and ensuring volunteers felt valued and heard. Integrating volunteers into decision-making processes fostered a sense of ownership and alignment with broader program goals, leading to more effective and engaged participation. Volunteers who understood the long-term impact of their contributions were more likely to remain committed and proactive in future crisis responses. These lessons have informed a more strategic, resilient approach to volunteer engagement, ensuring that public institutions, private-sector actors, and civil society organizations can work together more effectively in future emergencies.

The Social Construction of the Target Population: Reaching the End Beneficiaries

In-depth interviews with officials and operators from municipal social services, Caritas Poland, and the Food Bank Federation—who collaborated with the Municipality of Warsaw during the COVID-19 crisis—revealed that the official income criterion for selecting food assistance beneficiaries, while unchanged during the pandemic, proved insufficient. Respondents highlighted the need to incorporate additional criteria to identify and address the population's diverse needs more effectively. Street-level bureaucrats (SLBs) and NGO operators involved in the Food Assistance Program (FAP) also reported significant shifts in the demographics of individuals seeking assistance during the pandemic. Before COVID-19, beneficiaries included a broad range of individuals, such as the elderly, children, the sick, and people with disabilities. However, the pandemic triggered a noticeable shift in vulnerability, particularly among the elderly, who emerged as a group in greater need of support. This change is primarily attributed to the aging population and the fact that individuals over 65 are typically not engaged in the labor market.

Additionally, the Polish pension system needs help to provide adequate support, especially during periods of economic strain. As a result, elderly individuals began to receive more assistance than they had before the onset of the pandemic. This shift underscores the growing need for new criteria for recognition of social target recognition and a more nuanced and responsive approach to food assistance that accounts for these changing patterns of vulnerability.

During the crises, Poland's charitable food assistance system remained unchanged, with Caritas overseeing various forms of food aid. In Poland, there were 700 support centers funded by the European FEAD. Similarly to other countries, food procurement was managed by a government agency and distributed through regional centers. It is important to note that the centers involved in the FEAD-Caritas projects differ from the Food Banks. The FEAD program (2014–2020) ended in 2023, and new funding was anticipated with the new program adopted by the European Union (Interview 2, Caritas).

The ability to tailor food aid interventions to individual cases becomes difficult owing to the large influx of people during a crisis. Urban administration, social workers, and NGOs in Warsaw have turned to digital food stamps and cards to address this challenge. These tools have proven instrumental in streamlining food assistance efforts, particularly in the face of a rising demand from individuals seeking aid. While digital food stamps and cards empower citizens and households by offering them more autonomy in their food choices, they also facilitate the standardization of tools and methods for food assistance in critical conditions. However, as the number of beneficiaries grows, ensuring precise targeting and efficient delivery becomes increasingly complex. In the face of the immense challenge of managing large populations during crises, humanitarian organizations and NGOs that received funding and partnered with the Municipality during the COVID-19 pandemic and the Russian-Ukrainian war were tasked with developing robust systems for analyzing and selecting applicants for food aid. These organizations, leveraging their accumulated experience and employing straightforward assessment forms, were responsible for certifying the level of need among individuals and households requesting assistance. Once the state of need was verified, aid distribution was carried out through a coordinated network of volunteers, ensuring that resources reached those most in need efficiently and in a timely manner.

Both public and private street-level operators had to adjust their criteria for identifying target populations in response to evolving crises, shifting financial resources, and urgent needs. This process required adopting more flexible and responsive assessment tools that accounted for the unique vulnerabilities of affected groups. A key lesson from these adaptations was the need to reduce conditionality in accessing food aid. Strict eligibility criteria often delayed or restricted assistance to those in urgent need. By loosening these conditions, aid providers were able to reach a broader and more diverse group of beneficiaries, thus ensuring that support was not withheld owing to bureaucratic hurdles. This shift toward greater inclusivity and responsiveness improved the efficiency and fairness of aid distribution, enabling resources to be allocated where they were most needed. It also underscored the importance of balancing accountability with accessibility, ensuring that crisis responses remained both effective and equitable. However, a critical challenge that emerged was the disparity in treatment between different immigrant groups. While humanitarian aid for Ukrainian refugees was swiftly mobilized with broad public and institutional support, other migrant populations—particularly those from the Middle East, Africa, and Asia—often faced greater barriers to accessing assistance, including stricter eligibility criteria and less social acceptance. This uneven approach highlighted deeper structural biases in crisis response frameworks, raising concerns about the consistency and fairness of humanitarian aid distribution.

Engaging in Adaptation Processes through Monitoring and Evaluation

During the COVID-19 pandemic, the demand for food assistance surged, particularly from individuals who had never sought help or lost their jobs. Digitization played a key role in managing the increased demand, streamlining procedures, and monitoring the evolving needs of both municipal operators and private partners.

Significant changes were made to the home food assistance process for those lacking self-sufficiency. Initially outsourced to an external company, the service evolved from municipal caregivers cooking and delivering food to a more efficient QR code system. This system facilitated secure digital recognition of beneficiaries and food service operators, allowing for precise tracking of delivery times and ensuring minimal contact. The QR code system remained in place even after the pandemic as direct food deliveries were phased back through local centers.

Outsourcing home food assistance services during the pandemic allowed healthcare personnel to focus on direct medical care, optimizing resources, and improving the efficiency of food distribution. The expansion of electronic payment systems in the Urban Food Assistance Program (FAP) further enhanced monitoring, ensuring compliance with guidelines. The digitization of social services also led to digital food stamps, thus simplifying access to food assistance while enhancing safety measures. Another important innovation was the introduction of food cards for elderly individuals, implemented through partnerships with private companies like the Geronimo Martins Foundation, which operates 3000 stores across Poland. Caritas, a key NGO, received an annual grant from the foundation to support elderly aid initiatives. Volunteers identified beneficiaries, who then completed a financial assessment form. Implementing food cards gave elderly individuals more choice and dignity, with electronic payment systems allowing for real-time monitoring of purchase quality.

Additionally, the cards automatically restrict purchases of items such as alcohol and luxury goods, ensuring the aid supports healthy and responsible consumption. These digital innovations not only improved the efficiency of food assistance but also ensured greater transparency, accountability, and autonomy for beneficiaries. By integrating digital tools, the city and NGOs could better address the needs of vulnerable populations, providing more effective and streamlined support during the pandemic and beyond.

Turning Crisis Experience Insights into more Innovative Policies

After enduring three consecutive crises, SLBs and NGO operators in Warsaw have become better equipped to handle new challenges, improving aid interventions, procedures, and digital tools such as food stamps. However, long-term reliance on volunteers remains a challenge owing to potential fatigue. Despite efforts to reduce food waste and improve social initiatives, Warsaw has struggled to establish a unified food policy, with initiatives remaining fragmented and primarily focused on pandemic- and war-related emergencies. It highlighted the need for a more integrated, sustainable food policy that combines food aid, waste reduction, and broader social initiatives. The poly-crises also revealed new social issues, such as a rise in domestic violence and youth psychological distress, which require urgent policy attention. The increasing suicide rates, mainly since the COVID-19 pandemic, have prompted a review of data and a rethinking of support strategies. These challenges present opportunities to improve social policies, focusing on data-driven decision-making and inter-departmental cooperation. The Urban Food Assistance Program (FAP) has been beneficial for residents and Ukrainian refugees. However, its impact on other migrant groups, including asylum seekers and refugees from other regions, has been limited. These groups face unique challenges like language barriers, cultural differences, and legal issues, requiring more specialized support. Tailored programs and stronger collaboration between government, NGOs, and international organizations must ensure all migrant populations receive adequate assistance during crises (Interviews 7, 8).

Embedding Innovative Solutions into Routine Policy Frameworks

The municipal food assistance programs developed during the pandemic introduced a range of innovative tools that have since been integrated into routine support systems for elderly, vulnerable, and homeless populations, especially during the refugee crisis. Some preexisting services were expanded, while new initiatives were introduced to better address the needs of these communities. Among the innovations were city buses offering free food and assistance to homeless individuals and directing them to shelters, canteens, and dormitories. Additionally, kitchens for ready meals, food bags, and food stamps became more widely available. New assistance methods were also implemented, such as cash food aid, food cards, and food packages delivered by mobile consultancy centers.

In Warsaw, the 'homeless bus' service, which had been operational before the pandemic, initially provided food to those experiencing homelessness. However, this service was scaled up significantly during the lockdown period, transforming the bus into a permanent mobile food distribution center. Small portions of food were prepared for immediate consumption, attracting long queues as the demand for assistance surged.

The team responsible for food policy at the Warsaw Municipality believes that their innovative approach could serve as a model for other cities, helping them to better prepare for crises and improve their food systems. 'We strongly believe that food is a pivotal issue for the future, and the city can play a transformative role,' they stated (Interview 1). The experience gained during these crises has been shared on a European level, leading to Warsaw's participation in the 'Food Trails' project. This initiative facilitated the establishment of Warsaw's first full-scale living lab focused on food policy and food assistance, serving as a platform for collaborative creation and testing of innovative food solutions.

The ambition is for public–private collaboration, which includes diverse stakeholders such as universities, to evolve into a structured food governance system based on the Milan model capable of addressing future challenges (Interview 1). While there is broad recognition in Warsaw of the importance of food policy, the city currently needs an institutionalized, integrated policy approach. The existing food living lab aims to close this gap by constantly monitoring the needs of residents and stakeholders, ultimately committing to a more resilient and sustainable food system.

The Warsaw case reflects broader trends observed across Polish cities, both large and small, in navigating overlapping crises. Similar tools and strategies were adopted elsewhere, demonstrating the adaptability of collaborative approaches to varying contexts. For instance, Kraków—Poland's second-largest city with around 800,000 inhabitants—offers a compelling example of effective public–private coordination in food aid policies, particularly in supporting war refugees. At the heart of Kraków's efforts was the relocation information point at the Central Station, which served as a hub for refugee registration, accommodation, and relocation. This initiative was complemented by services from the Polish Scouting and Guiding Association (ZHP) and the Maltese Aid Service, who provided rest areas, catering, and first aid.

The TAURON Arena Kraków also operated a registration desk from mid-March to June 2022, handling over 31,000 PESEL applications despite significant overcrowding and operational challenges. The collaborative nature of this initiative—bringing together local government units, NGOs, and international organizations like the UNHCR (operating a *Blue Dot* point)—set it apart from efforts in other medium-sized Polish cities. However, the overwhelming influx of refugees occasionally strained grassroots initiatives and created resource management and space allocation tensions. Similar food assistance efforts extended to smaller cities, towns, and villages near Poland's borders. These initiatives were largely successful owing to robust community involvement, including the active participation of local officials. Interestingly, inter-institutional coordination was often informal, relying on personal relationships rather than structured frameworks, while vague guidelines from the central government posed additional challenges. In response to the influx, collective reception centers were established across cities and provinces, accommodating thousands of refugees.

Nevertheless, many refugees—particularly those arriving after Russia's annexation of Crimea in 2014—found shelter in private homes, supported by Polish and Ukrainian families. A Union of Polish Metropolises report revealed that 67% of residents in Poland's 12 largest cities participated in initiatives to support Ukrainian refugees. The most common forms of aid included food donations (59%), in-kind assistance such as clothing or furnishings (44%), and monetary contributions or participation in fundraisers

(32%). Kraków's grassroots efforts mirrored those witnessed in Warsaw and other cities. Independent warehouses managed by organizations such as the Nidaros Foundation and the Multicultural Center coordinated the collection and distribution of necessities. In Warsaw, specialized bodies such as the SOS Food Bank played a pivotal role in projects such as *'Survive with Dignity. Food for People from Ukraine,'* which focused on collecting non-perishable food through retail partnerships. Similar efforts were replicated in Rzeszów, where the Podkarpackie Food Bank led the response, and in Przemyśl, where the Salvation Army managed food collection and distribution.

While Warsaw is a prominent example of coordinated efforts at scale, Kraków and smaller towns are proof that successful models can adapt to different contexts, relying on institutional frameworks and grassroots support to address urgent needs.

4.5 Milan's Urban Food Security Program: Rising to Turbulent Times

The adoption of Milan's Municipal Food Policy traces back to Expo 2015 ('Feed the Planet') and its pivotal event: the approval of the Milan Urban Food Policy Pact (MUFPP). This pact, launched under the leadership of the mayor of Milan, established a network of cities committed to creating socially and environmentally sustainable food systems. The MUFPP marked the first international collaborative effort among cities to share best practices on food security. It also incentivized signatory cities to adopt individual food policies, aligning urban food governance objectives and implementation plans (Candel, 2020). The pact encouraged European cities to develop tailored, context-specific food policies, drawing from one another's experiences. However, despite these efforts, Milan's urban policy agenda did not prioritize food poverty and vulnerability until the COVID-19 pandemic in 2020. Before the pandemic, Milan's food security initiatives primarily focused on 'sustainable diet and nutrition,' in line with the MUFPP narrative. The city's food security efforts were mainly carried out by third-sector organizations that managed FEAD (European Fund for Aid to the Most Deprived) resources and redistributed surplus food. With the onset of the pandemic, Milan took new institutional steps to address food poverty, introducing direct food distribution and income support measures, such as the Food Aid Device and Shopping Vouchers.

Since its establishment, the Municipal Food Policy Office has been primarily recognized as a body focused on project development and community engagement (Calori, 2022). Integrated within the Mayor's Office, it works across sectors to create initiatives without duplicating the responsibilities of existing municipal departments, such as those handling school canteens, agriculture, waste management, and trade. Instead, the Food Policy Office serves as a technical support body, assisting these departments in developing innovative projects and fostering cross-sectoral integration to address food-related issues more systemically. A key aspect of the Office's evolution is its strong connection between local actions and international collaborations. European projects secured by the Office have provided additional resources, enabled gradual staff expansion, and enhanced its operational capacity while maintaining its technical role. Initially, the Office's primary focus was combating food waste, with numerous projects aimed at schools, markets, and public and private catering sectors. Over time, however, other areas, such as education, urban and peri-urban agriculture, local sourcing for school canteens, and integrating food systems with urban regeneration, gained greater prominence. Following Mayor Sala's re-election and recognizing the Office's significant work during the pandemic, a major reorganization of municipal offices began in the autumn of 2021. This reorganization led to substantial changes and a repositioning of the Food Policy Office. In April 2022, the Food Policy Area was created, overseen by a newly appointed director and structured into six distinct units. These units now manage not only administrative tasks but also agriculture, the coordination of Milano Ristorazione (school canteens), food poverty, partnerships, and the Milan Urban Food Policy Pact (MUFPP), with a strong focus on knowledge transfer.

Breaking out of Milan's Food Policy Comfort Zone

Milan's food policy stands out for its long-term and evolving approach to collaborative governance. It began in 2014 with city-led consultations involving diverse stakeholders and informal groups to shape the policy framework (Calori & Magarini, 2015). According to the Director of the Municipal Food Policy Area, the COVID-19 pandemic and the subsequent food emergency management served as a pivotal moment, expanding the capacity of local public actors to broaden the scope of the city's food policies. The crisis heightened the emphasis on food security, integrating it into broader policy discussions and enhancing Milan's ability to lead a wide range of stakeholders within its urban and peri-urban food systems. This evolution highlights the transformative impact of external shocks, such as the pandemic, in compelling policymakers and partners to rethink strategies and adopt more comprehensive crisis management frameworks. These new approaches address immediate needs, strengthen resilience, and foster a more integrated food system for the future. As a result, Milan's food policy has significantly adapted, incorporating improved coordination mechanisms, enhanced stakeholder engagement, and data-driven strategies to navigate future challenges better while continuing to lead in sustainable urban food governance.

Reinforcing Trust through Collaborative Food Security Governance

Looking at the extensive narrative surrounding Milan's Food Policy, one might assume that the pandemic contributed little to improving its well-established collaborative and inclusive governance. Before COVID-19, the city's food policy governance operated within a highly institutionalized collaborative framework characterized by shared responsibility among key stakeholders. These included the Cariplo Foundation, which provided direct coordination support, the Municipal Food Policy Office within the Mayor's Cabinet, and a wide range of public and private entities: Milano Ristorazione, SogeMi, AMAT, MM, Politecnico di Milano, Assolombarda, ATS Milano, the Regional School Office, Parco Agricolo Sud Milano, Agricultural Districts, Slow Food Italia, and Communities of Practice encompassing public, private, and civil society organizations. During 2018 and 2019, this coordinated governance became more robust and efficient, with learning primarily derived from Communities of Practice, project planners, and initiatives supported by European funding (Magarini, 2022). However, beginning in March 2020, the pandemic amplified the collaborative experience developed over previous years and created opportunities to address emerging food needs and challenges arising from its effects. The active engagement of public and private sectors, civil society, and academic institutions enabled Milan to adopt a holistic strategy that tackled the multifaceted issues of urban food security and sustainability. Milan's experience underscores the importance of long-term commitment and strategic planning in building inclusive, trust-based food policies that adapt to evolving circumstances. The strategic partnership with the Cariplo

Foundation was pivotal, with frequent interactions between the Municipality, the Foundation, NGOs, associations, and other private actors facilitating knowledge sharing, the integration of approaches and tools, and the removal of barriers. After an initial adjustment period, these interactions consolidated mutual trust, enabling the gradual expansion of the partnership to include a broader network of actors involved in co-creating food policy actions. The synergistic relationship between the Municipality and the Cariplo Foundation demonstrates the value of strategic partnerships in urban governance. Their collaboration fostered a cohesive strategy, enhanced by the credibility and trust they built together. This trust allowed for the mobilization of resources and the engagement of a wide range of stakeholders, including public institutions, private entities, nonprofits, and community organizations. By embracing an inclusive approach, the city ensured that diverse perspectives and bodies of expertise contributed to its food policies.

Additionally, the municipal focus on a 'public value' narrative and shared goals mitigated potential conflicts, creating a unified framework that is better equipped to address complex challenges. A striking example of this collaboration emerged during the pandemic. Between March 9th and 15th, 2020, Milan's social policies, food policy, and civil protection structures jointly designed an integrated Food Aid Device. This initiative centralized the procurement and distribution of food in close partnership with municipal companies, FEAD leaders (the Lombardy Food Bank, the Italian Red Cross, Milan Committee, and Caritas Ambrosiana), and the IBVA social market, with additional support from the Cariplo Foundation's QuBì Program. From March 16th to June 30th, 2020, the Food Aid Device operated during the acute lockdown phase, centralizing supply chains and managing weekly food distributions. All the joint efforts increased the services offered to new families referred by the city. Volunteers played an essential role in bridging gaps between institutional responses and community needs. Their on-the-ground presence provided critical insights into the challenges faced by vulnerable groups, ensuring interventions were timely and targeted. Collaboration between public institutions and private entities further amplified these efforts, pooling diverse resources and expertise to enhance the response's effectiveness significantly. The pandemic underscored the value of a robust public–private volunteer network in addressing crises, demonstrating that collective action can bolster resilience and adaptability in urban food systems.

New Emerging Vulnerable Populations

With the onset of the pandemic, the Milan Food Office contributed to reorienting many municipal public services to address the challenges posed by COVID-19. Among the most pressing concerns were ensuring the uninterrupted distribution of food aid during the initial lockdown period and scaling up operations to meet the surge in food insecurity. This effort was spearheaded by a collaborative framework involving the Food Policy Office, the Directorate of Social Policies, and the Civil Defense of the city of Milan, working within the Municipal Operations Center (COC)—the city's hub for emergency management. Beginning in March 2020, officials from these three bodies convened regularly at the COC to strategize and coordinate Milan's pandemic response. The city adopted a comprehensive approach to ensure that all residents, especially the most vulnerable, had access to essential resources, particularly food. The coordination among the Food Policy Office, the Directorate of Social Policies, and the Civil Protection proved instrumental in streamlining resources and efforts. The COC facilitated real-time collaboration and enabled the swift implementation of measures to address the immediate and longer-term impacts of the crisis. Milan's preexisting culture of collaboration proved invaluable in rapidly revising strategies and deploying substantive and procedural tools. On March 29th, 2020, the National Head of Civil Defense issued Ordinance 658/2020, which allocated resources for acquiring shopping vouchers for food and other essential products. His integration, alongside the Food Aid Device, created a versatile response mechanism tailored to address the diverse needs of Milan's residents. By combining direct food distribution with shopping vouchers, the city provided a range of effective solutions for various citizen groups. Years of experience and trust among stakeholders facilitated a swift and coordinated response, allowing the city's technical structures to adapt quickly to new challenges. This adaptability was crucial in addressing the rising food insecurity and other issues brought on by the pandemic, showcasing Milan's ability to innovate and meet its community's needs in a crisis.

Looking ahead, the lessons learned during the pandemic are shaping the future direction of Milan's food policy. The focus is developing more resilient and adaptable food systems capable of withstanding future shocks. It includes exploring innovative solutions to meet evolving food needs, strengthening local food networks, and continuing to foster strong partnerships across all sectors. The

> *aim is to build a sustainable and inclusive food system that ensures food security for all residents, regardless of circumstances.* (Interview 1.2)

Integration at both functional and structural levels—across policy domains, programs, public entities, private actors, and stakeholders—was pivotal in adapting and reorienting Milan's food security policy into a more comprehensive and systemic urban food framework. This integration streamlined municipal efforts and resources, ensuring coordinated interventions that effectively addressed immediate food insecurity while bolstering the urban food system's long-term resilience. From 2020 to 2021, the city of Milan's urban food actions were focused on addressing food poverty and emerging vulnerabilities. A cornerstone of this response was the launch of the *Milano Aiuta* task force, a city-led initiative that coordinated emergency measures across public and private sectors. The Food Policy Office played a key role in initiating the Food Aid Framework. During 2020 alone, the city provided food aid—via packages and shopping vouchers—to approximately 100,000 residents. This effort mobilized municipal human, financial, real estate, and relational resources while fostering a robust network of private and social–private actors, establishing a new operational area for Milan's future food policy. The *Milano Aiuta* task force embodied a proactive and integrated approach to crisis management by uniting diverse stakeholders for a cohesive response to food insecurity. Through 10 Temporary Hubs staffed by municipal employees, volunteers, and third-sector operators, the initiative centralized food supply chains, facilitated the assembly of food packages for families, and ensured swift and effective distribution. This effort leveraged the strengths of public–private collaboration, with the Food Policy Office aligning activities with broader food policy goals to promote sustainability and resilience. To address the heightened needs during the second lockdown period in the winter of 2020, Milan's local government extended additional support to third-sector organizations, redesigning the Food Aid Device to involve volunteers and third-party entities in its management directly. Municipal companies were instrumental in facilitating logistics; notably, the Urban Mobility Company (AMAT) played a crucial role in managing the extraordinary home delivery plan supported by the Food Bank and Milan's Red Cross. Building on the success of the *Milano Aiuta* initiative, the Municipality formalized partnerships with third-sector organizations to enhance their capacity to address local food aid needs. The Food Aid Program was refinanced through calls for tenders in 2021 and 2022, enabling third-sector entities to procure and distribute food to vulnerable populations. Tender evaluations prioritized efficiency, collaboration, food quality, and clear beneficiary identification criteria. Third-sector entities ensure the quality of foodstuffs and establish effective criteria for identifying beneficiaries. The Councilor for Welfare emphasized the program's dual significance: combating poverty and marginalization while equipping social workers with a critical resource to identify and assist vulnerable households. The initiative selected up to nine organizations, with funds covering up to 90% of the project budget. Each project was expected to secure approximately 40–50 tons of food, translating into about 5000 food parcels per project and a total of 45,000 parcels distributed across the city. The program complemented ongoing efforts by third sector entities accredited under the European Fund for Aid to the Most Deprived (FEAD), which includes 193 organizations in Milan. Beneficiaries comprised existing aid recipients and new individuals identified through local networks or municipal referrals. The Welfare and Health Directorate, in partnership with the Food Policy Area, oversees the program's implementation, ensuring accountability and alignment with the city's broader welfare goals.

Boosting Data Drives Governance to Facilitate Policy Reuse

The Milan Food Aid Device provided critical support to over 20,700 individuals, including 7139 children and 514 infants, as well as more than 6300 families in need, distributing a total of over 616 tons of food across nearly 50,000 aid deliveries (Magarini, 2022). This large-scale initiative was instrumental in mitigating food insecurity during a particularly challenging time. The rigorous monitoring and data collection efforts of the Food Policy Office underpinned its success. By systematically tracking all aspects of aid distribution, the Office ensured the production of clear, accurate, and timely reports, which proved essential for maintaining transparency and accountability. These reports were not merely internal documents, they were also publicly presented during City Council hearings, reflecting Milan's commitment to openness and fostering trust among its residents. This transparency highlighted the city's proactive approach to food insecurity and reinforced public confidence in its crisis response efforts. Furthermore, the success of the Food Aid Device established a robust framework for future emergency responses. The monitoring practices, data-driven insights, and well-coordinated networks of partners developed during this initiative have equipped Milan with the tools and strategies needed to address similar challenges in the future. The experience gained from the Food Aid Device has significantly contributed to the city's ongoing food policy efforts, laying a strong foundation for building a more resilient and food-secure urban environment. Milan's food policy underwent a major transformation during the pandemic, becoming deeply integrated with the city's broader social policies. This integration not only spurred innovation within the food policy framework but also enhanced the capabilities of the public welfare system. Starting in September 2020, combating food poverty became a central focus, reflecting a shift in priorities and recognizing food security as a vital component of social welfare. Rigorous monitoring activities, encompassing both the supply and demand sides of food distribution, were critical to this process. These efforts ensured that initiatives were responsive to immediate needs and sustainable in the long term. Through continuous data collection and analysis, Milan could assess the impact of its programs, make data-informed adjustments, and allocate resources effectively. Integrating food and social policies allowed for a more holistic approach to addressing poverty and inequality, enabling the city to tackle the root causes of food insecurity while improving the resilience of its public welfare system. The pandemic underscored the need for flexible and responsive data collection systems. Traditional data-gathering methods often fall short of capturing vulnerable populations' dynamic and nuanced needs, especially during crises. Therefore, developing integrated monitoring systems that can quickly adapt to changing circumstances is critical. The systems must be capable of collecting real-time data, analyzing trends, and informing decision-making processes to ensure that support is directed where it is most needed. Looking ahead, the lessons learned during the pandemic highlight the importance of institutionalizing these practices. Establishing permanent structures and processes for collaboration, data integration, and adaptive management will enhance Milan's capacity to respond to future crises. By fostering a culture of continuous learning and improvement, Milan can strengthen its food security governance and ensure that all residents are supported, even in the face of unforeseen challenges. The continuous emphasis on anti-food poverty initiatives is a strategic pivot toward establishing lasting and resilient solutions within Milan's food security framework. The city's sustained collaboration with third sector organizations is central to this approach, which remains a cornerstone of its strategy. These entities provide invaluable expertise and resources, particularly identifying new fragilities and managing and distributing FEAD aid and surplus food. By integrating these partnerships into a systematic and organized framework, Milan ensures that its efforts are effective in addressing immediate needs and that they are sustainable in the long term. This approach aims to build a robust and coordinated network capable of addressing food insecurity in a more efficient and impactful way. Importantly, the city's strategy transcended short-term relief by incorporating long-term initiatives, such as the development of Neighborhood Hubs. These hubs remain central nodes of food distribution and community support today, fostering a more interconnected and responsive urban food system. Furthermore, Milan's food policy extends across multiple sectors, including targeted interventions within the wholesale market to streamline food distribution and improve accessibility for vulnerable populations. Developing cross-sectoral tools and resources reinforces the comprehensiveness and adaptability of the city's approach to food security. This flexibility enables Milan to respond effectively to future challenges, including climate change and other systemic crises. The lessons learned during the pandemic have profoundly shaped Milan's food policy, driving a forward-looking approach that prioritizes integration, resilience, and innovation. This evolving framework underscores the city's commitment to building an inclusive and sustainable urban food system that meets current and future needs.

4.6 What the Case Studies Reveal

The contrasting governance contexts of Milan and Warsaw significantly shaped their responses to food insecurity during crises, revealing key differences in institutional maturity, strategic approaches, and learning processes. Milan's structured and proactive model leveraged preexisting governance mechanisms, a dedicated food policy framework, and collaborative networks, allowing for coordinated, long-term planning. In contrast, Warsaw's centralized and reactive approach prioritized efficiency but relied on ad hoc responses, limiting grassroots engagement and institutional learning. The main differences between the strategies, the learning processes, and the use of knowledge during the post-crisis period in the two contexts are as follows.

Strategic Collaboration and Policy Flexibility. Milan's decentralized welfare system and autonomous institutional settings enabled a proactive and inclusive response that emphasized co-creation and cross-sector collaboration. The city integrated third-sector organizations experienced in FEAD (Fund for European Aid to the Most Deprived) projects, fostering an adaptable, nuanced approach to food security. This flexibility enabled Milan to institutionalize collaborative networks, incorporate crisis lessons into policy, and expand its Food Policy Office to address broader food security dimensions after the pandemic. Warsaw, on the other hand, followed a centralized governance model in which municipal authorities collaborated closely with the central government. Large charitable organizations, such as food banks and Caritas, worked mainly with national institutions and beneficiaries rather than engaging with city authorities or local grassroots organizations. This approach facilitated rapid decision-making and resource allocation, ensuring immediate relief but limiting opportunities for systemic transformation and stakeholder-driven innovation.

Institutional and Operational Learning Processes. The learning processes underlying the Food Aid Programs (FAP) in both cities highlight notable differences in knowledge acquisition and transformation. Milan's decentralized and collaborative model fostered institutionalized learning, enabling stakeholders to co-design and co-manage food security interventions. The inclusive approach enhanced adaptability and responsiveness, ensuring interventions aligned with the evolving needs of vulnerable populations. Conversely, Warsaw's learning process was more contingent and ad hoc (Capano et al., 2019). The city's centralized framework prioritized rapid mobilization through established networks, effectively addressing immediate needs but lacking iterative, co-creative learning processes seen in Milan. Warsaw's reliance on a small number of large FEAD organizations ensured operational continuity and efficiency but restricted broader engagement and innovation opportunities.

Knowledge Reuse and Long-Term Impact. The extent to which each city institutionalized crisis-related knowledge further underscores their contrasting approaches. In Warsaw, the reuse of existing frameworks and partnerships was evident during the Ukrainian refugee crisis, where established FEAD networks helped manage new socio-economic pressures. While this showed the value of continuity, learning and adaptation were confined to a limited set of actors, preventing broader institutional transformation. In contrast, Milan catalyzed long-term institutional change, embedding food security initiatives within its governance structure. Its inclusive, decentralized approach ensured that knowledge was widely disseminated across various stakeholders, allowing for policy innovation and systemic resilience. By prioritizing co-creation and flexibility, Milan laid the groundwork for a robust, future-oriented food security governance model.

Governance Models and Crisis Resilience. The governance models of Milan and Warsaw illustrate how institutional configurations shape crisis responses, adaptability, and long-term resilience. Warsaw's centralized model ensured efficient resource mobilization but lacked grassroots engagement and institutional learning, while Milan's decentralized approach fostered community participation and street-level policy innovation, requiring strong coordination to maintain consistency. These contrasting experiences offer valuable lessons for cities seeking to enhance food security governance in times of crisis. Milan's collaborative, innovation-driven resilience and Warsaw's efficiency-oriented, centralized model highlight the importance of institutional maturity, resource availability, multi-stakeholder collaboration, and the ability to integrate and reuse knowledge in responding to poly-crises. Moving forward, balancing immediate relief with long-term inclusivity and systemic transformation will be key to building sustainable and equitable food security governance frameworks.

Notes

1. Municipal Volunteering Portal, www.ochotnicy.waw.pl, (December 6th, 2024)
2. Warsaw currently enjoys the status of a city with county rights. From 1994 to 2002, Warsaw was divided into several independent municipalities, but the model proved ineffective, so the formula of the single municipality divided into several auxiliary units districts was adopted. The division of tasks between Warsaw and the districts is not specifically defined on a statutory level, so the Warsaw City Council decides on which tasks will be assigned to the districts.

References

Borràs-Pentinat, S. (2020). Environmental refugees: Reshaping the borders of migration in the EU. In *The European Union and global environmental protection* (pp. 109–131). Routledge.

Bouchet, C., & Duvoux, N. (2023). Post-COVID-19 poverty in France: Magnitude, manifestations and actors from nine case studies. *International Journal of Sociology and Social Policy, 43*(9/10), 823–837.

Busetti, S., & Pace, N. (2023). *Food loss and waste policy*. Routledge.

Calori, A. (2022). La Food Policy di Milano: Preistoria e evoluzione organizzativa. *Rel Cibo, 1*(1), 22–39.

Calori, A., & Magarini, A. (2015). The role of informal food networks. Notes from the experience of Milan toward food policy councils. *Economia e Sostenibilita*, 1–9. https://www.fao.org/fileadmin/templates/ags/docs/MUFN/CALL_FILES_EXPERT_2015/CFP1-11_Full_Paper.pdf.

Candel, J. J. (2020). What's on the menu? A global assessment of MUFPP signatory cities' food strategies. *Agroecology and Sustainable Food Systems, 44*(7), 919–946.

Capano, G., Howlett, M., & Ramesh, M. (2019). Disentangling the mechanistic chain for better policy design. In Making policies work (pp. 2–13). Edward Elgar Publishing. Retrieved June 14, 2023, from https://www.elgaronline.com/display/edcoll/9781788118187/9781788118187.00008.xml.

Cheek, N. N., Reutskaja, E., & Schwartz, B. (2022). Balancing the freedom–security trade-off during crises and disasters. *Perspectives on Psychological Science, 17*(4), 1024–1049.

Crayne, M. P. (2020). The traumatic impact of job loss and job search in the aftermath of COVID-19. *Psychological Trauma: Theory, Research, Practice, and Policy, 12*(S1), S180.

Crush, J., & Fayne, B. (2010). *The invisible crisis: Urban food security in Southern Africa* (Vol. 1). AFSUN.

Davis, O., & Geiger, B. B. (2017). Did food insecurity rise across Europe after the 2008 crisis? An analysis across welfare regimes. *Social Policy and Society, 16*(3), 343–360.

Doernberg, A., Horn, P., Zasada, I., & Piorr, A. (2019). Urban food policies in German city regions: An overview of key players and policy instruments. *Food Policy, 89*, 101782.

Eurocity. (2021). *Food aid in European cities.* Working Group Food Policies, March, Brussels, 1–31.

FAO. (2019). *FAO framework for the urban food agenda.* FAO Knowledge Repository website. Retrieved November 17, 2024, from https://openknowledge.fao.org/handle/20.500.14283/ca3151en

FAO. (2023). *FAOSTAT: Suite of food security indicators.* FAO Knowledge Repository website. Retrieved December 16, 2024, from https://openknowledge.fao.org/server/api/core/bitstreams/f1ee0c49-04e7-43df-9b83-6820f4f37ca9/content/state-food-security-and-nutrition-2023/food-security-nutrition-indicators.html

Garthwite, K. (2015). *Hunger pains. Life inside foodbank Britain.* Policy Press.

Filippini, R., Mazzocchi, C., & Corsi, S. (2019). The contribution of urban food policies toward food security in developing and developed countries: A network analysis approach. *Sustainable Cities and Society, 47*, 101506.

Goedemé, T., Penne, T., Swedrup, O., Van den Bosch, K., & Storms, B. (2019). *Is there common ground for defining a decent social minimum in Europe?* Oxford University Research Archive.

González-Azcárate, M., Cruz-Maceín, J. L., Bardají, I., & García-Rodríguez, A. (2023). Local food policies from a city-region approach: Fostering the SFSCs in the region of Madrid. *Cities, 133*, 104158.

Jablonski, B. B. R., Casnovsky, J., Clark, J. K., Cleary, R., Feingold, B., Freedman, D., Gray, S., Romeiko, X., Olabisi, L. S., Torres, M., van den Berg, A. E., Walsh, C., & Wentworth, C. (2021). Emergency food provision for children and families during the COVID-19 pandemic: Examples from five US cities. *Applied Economic Perspectives and Policy, 43*(1), 169–184.

Konstantinidis, C. (2024). Food insecurity, austerity, and household food production in Greece, 2009–2014. *Review of Social Economy, 82*(4), 497–535.

Kovács, E. K., Bachórz, A., Bunzl, N., Mincyte, D., Parasecoli, F., Piras, S., & Varga, M. (2022). The war in Ukraine and food security in Eastern Europe. *Gastronomica: The Journal of Food and Culture, 22*(3), 1–7. Retrieved December 2, 2024, from https://online.ucpress.edu/gastronomica/article-abstract/22/3/1/192112.

Kulesza, M., & SzeŚciŁo, D. (2012). Local government in Poland. In A. Manuel-Moreno (Ed.), *Local government in the member states of the European Union: A comparative legal perspective* (pp. 485–504). INAP.

Lam, D. P. M., Jiménez-Aceituno, A., Guerrero Lara, L., Sellberg, M. M., Norström, A. V., Moore, M. L., Peterson, G. D., & Olsson, P. (2022). Amplifying actions for food system transformation: Insights from the Stockholm region. *Sustainability Science, 17*(6), 2379–2395. https://doi.org/10.1007/s11625-022-01154-7

Lambie-Mumford, H., & Dowler, E. (2015). Hunger, food charity and social policy–challenges faced by the emerging evidence base. *Social Policy and Society, 14*(3), 497–506.

Lambie-Mumford, H., & Silvasti, T. (Eds.). (2020). *The rise of food charity in Europe*. Policy Press.

Lang, T., Barling, D., & Caraher, M. (2009). *Food policy: Integrating health, environment and society*. OUP Oxford.

Loh, P., & Agyeman, J. (2019). Urban food sharing and the emerging Boston food solidarity economy. *Geoforum, 99*, 213–222.

Loopstra, R., Reeves, A., & Stuckler, D. (2015). Rising food insecurity in Europe. *Lancet, 385*(9982), 2041.

López Cifuentes, M., & Fiala, V. (2022). Covid-19 as a chance for more food democracy in European cities? The responses of actors within Vienna's urban food system to the pandemic, London. *Cities, 131*, 104041. https://doi.org/10.1016/j.cities.2022.104041

Magarini, A. (2022). Politiche del cibo e governance collaborativa nell'esperienza di Milano, prima e dopo la pandemia da COVID-19. Leadership facilitativa e istituzionalizzazione. *Rivista Italiana di Politiche Pubbliche, 17*(3), 401–422.

Maino, F., & De Tommaso, C. V. (2022). Fostering policy change in anti-poverty schemes in Italy: Still a long way to go. *Social Sciences, 11*(8), 327.

Mansfield, B., & Mendes, W. (2013). Municipal food strategies and integrated approaches to urban agriculture: Exploring three cases from the global north. *International Planning Studies, 18*(1), 37–60.

Matusz, P., & Pawlak, M. (2020). Multi-level governance of integration policy, role of the cities. *Comparison of Warsaw and Prague, Political Science Review, 4*, 23–41. https://www.ceeol.com/search/article-detail?id=923006

Moragues-Faus, A. (2021). The emergence of city food networks: Rescaling the impact of urban food policies. *Food Policy, 103*, 102107.

Moragues-Faus, A., & Battersby, J. (2021). Urban food policies for a sustainable and just future: Concepts and tools for a renewed agenda. *Food Policy, 103*, 102124.

Moragues-Faus, A., & Morgan, K. (2015). Reframing the foodscape: The emergent world of urban food policy. *Environment and Planning A: Economy and Space, 47*(7), 1558–1573.

Morgan, K. (2009). Feeding the city: The challenge of urban food planning. *International Planning Studies, 14*(4), 341–348.

Polman, D., & Bazzan, G. (2023). Governance tools for urban food system policy innovations in the Milano urban food policy pact. *European Urban and Regional Studies, 30*(4), 362–337.

Righettini, M. S., & Lizzi, R. (2020). Governance arrangements for transboundary issues: Lessons from the food waste programs of Italian regions. *Review of Policy Research, 37*(1), 115–134.

Schoneville, H. (2018). Poverty and the transformation of the welfare (state) arrangement. Food banks and the charity economy in Germany. *Social Work & Society, 16*(2). 1-9. ISSN 1613-8953. http://nbn-resolving.de/urn:nbn:de:hbz:464-sws-1583

Selke, S. (2017). The new economy of poverty. In *Austerity, community action, and the future of citizenship in Europe* (pp. 197–216). Policy Press.

Skordili, S. (2013). Economic crisis as a catalyst for food planning in Athens. *International Planning Studies, 18*(1), 129–141.

Sonnino, R. (2019). The cultural dynamics of urban food governance. *City, Culture and Society, 16*, 12–17.

Sonnino, R. (2023). Food system transformation: Urban perspectives. *Cities, 134*, 104164. Retrieved November 19, 2024, from https://www.sciencedirect.com/science/article/pii/S0264275122006035.

Stierand, P. (2012). Food policy councils: Recovering the local level in food policy. In *Sustainable food planning: Evolving theory and practice* (pp. 67–78). Wageningen Academic.

Van der Horst, H., Pascucci, S., & Bol, W. (2014). The "dark side" of food banks? Exploring emotional responses of food bank receivers in the Netherlands. *British Food Journal, 116*(9), 1506–1520.

Vara-Sánchez, I., Gallar-Hernández, D., García-García, L., Alonso, N. M., & Moragues-Faus, A. (2021). The co-production of urban food policies: Exploring the emergence of new governance spaces in three Spanish cities. *Food Policy, 103*, 102120.

Wach, D., & Pachocka, M. (2022). Polish cities and their experience in integration activities-the case of Warsaw. *Studia Europejskie-Studies in European Affairs, 26*(2), 89–105.

Zerbian, T., & De Luis Romero, E. (2023). The role of cities in good governance for food security: Lessons from Madrid's urban food strategy. *Territory, Politics, Governance, 11*(4), 794–812. https://doi.org/10.1080/21622671.2021.1873174

CHAPTER 5

Conclusions: Solving the Puzzle of Street-Level Learning During Crisis

The first chapter of this book introduced the research questions and the central argument that, in times of poly-crisis, the collaborative governance of food security can be achieved and enhanced through mechanisms that empower street-level bureaucrats (SLBs) to foster trust and generate learning from the ground up. These frontline actors, under the pressure of new demands, play a crucial role in shaping adaptive and more resilient decision-making, through which one can respond to the growing and complex needs of food security. The chapters of this book have shown how food security governance has evolved across different levels of government in response to successive crises. This transformation has increasingly relied on bottom-up processes of collaborative decision-making and learning—facilitated by the expanding role of public policies, programs, and the practitioners who implement them. This research has highlighted previously under-explored street-level dimensions of food security governance and opportunities to build shared learning, thus offering fresh insights into dynamics that remain insufficiently addressed in the current academic debate. This concluding chapter dwells on final reflections on the initial collaborative learning framework, which provided a valuable lens through which to examine how various crises—from economic downturns to pandemics and conflicts—have contributed to reshaping food aid governance across institutional levels.

M. S. Righettini, *The Governance of Food Security in Europe*, https://doi.org/10.1007/978-3-031-98083-1_5

5.1 Street-level Learning Under Pressure

This book has investigated the crucial role of SLBs and procedural tools in times of crisis for enabling collaborative learning and building trust throughout the processes of knowledge acquisition, translation, and dissemination. Research conducted in a multilevel governance perspective has revealed additional mechanisms that complement those already identified in the existing literature (see Table 1.2). The observed mechanisms offer valuable insights into the potential of collaborative learning and trust-building in food security governance at the street level, particularly under poly-crisis pressures.

Broadening the scope of food security: Consecutive crises have led European institutions to raising public attention on food security and redesign procedures and tools to deal with it. The traditional approach of addressing food security by relying on knowledge of surplus agricultural production and redistributing the excess through charity is no longer considered sufficient. Food insecurity has instead become a complex, multidimensional policy challenge that intersects with social, environmental, economic, and health issues. European institutions increasingly view food security as an essential element to be integrated into cohesion policies, recognizing its significance for promoting social inclusion, reducing geographical disparities, and supporting sustainable development. As reflected in key EU strategies—such as the European Green Deal, the Farm to Fork Strategy, and the Common Agricultural Policy (CAP)—food security is not only about ensuring access to sufficient and nutritious food but also about addressing more comprehensive and open public–private–people collaboration in response to crises. In this context, cohesion policies serve as vital instruments for enhancing multilevel and cross-sectoral collaboration to address vulnerable populations and community resilience, thereby fostering procedural innovation in local governance. In this vein, street-level bureaucrats have emerged as key players in the food security landscape. They increasingly take on proactive roles, implementing diverse tools to respond to urgent needs while laying the groundwork for long-term, sustainable solutions. However, these efforts are undermined at the local level, where fragmented administrative actions and limited awareness of the broader implications of food security impede effective collaboration. This challenge is especially pronounced in areas where local organizations face difficulties securing resources or aligning their initiatives with national policy priorities, as in the case of Warsaw.

Food insecurity as a 'fire alarm' of hidden vulnerabilities: The rise in food insecurity—intensified by overlapping crises such as economic downturns, the pandemic, wars, and environmental disruptions—has acted as a warning signal, revealing previously hidden or underestimated social vulnerabilities. These include precarious employment, housing instability, limited access to healthcare, and social isolation. The study has revealed the risks of overexposure to food vulnerability by specific social categories in urban environments, such as immigrants, aged people, and isolated people and children. Under pressure, local welfare systems have gained new insights into the evolving needs of their populations, thus enabling them to design and implement more targeted, responsive interventions for at-risk groups. As a result, food insecurity is increasingly seen not just because of crisis but also as a valuable indicator for identifying emerging forms of social exclusion and guiding adaptive policy responses.

Assuming public responsibility: From reactive to proactive food security programs: In poly-crises, a paradigm shift has meant moving from reliance on charities and third-sector organizations to a more proactive, institutionalized approach, where public institutions take the lead in food security governance. While civil society organizations and philanthropic initiatives remain essential, governments increasingly recognize food security as a public responsibility rather than a matter of discretionary aid. This transformation reflects a broader governance shift from delegation to coordination, where the public is not merely an enabler of third-sector interventions but rather an active architect of multi-agency public policies. Several key governance dynamics drive the shift. Instead of compensating for state inaction, charities are increasingly co-designing and co-managing food security programs in collaboration with public institutions.

Changing the narrative: from 'food poverty' to 'food vulnerability': Recognizing new beneficiaries requires a shift in understanding—moving from the traditional concept of *food poverty* to the broader, more nuanced framework of *food vulnerability*. While food poverty was traditionally defined as financial constraints that limit access to sufficient food, food vulnerability—especially in times of overlapping crises—captures a more complex reality. It reflects the interplay of factors such as access to material support, food quality, education, health, and social inclusion. This evolving perspective recognizes that food insecurity is not solely a result of economic hardship but is instead shaped by systemic inequalities, environmental pressures, and structural barriers that affect individuals and communities differently. During crises such as the COVID-19 pandemic and the Ukrainian refugee influx, national governments recognized the diverse vulnerabilities of different population groups, including migrants, elderly individuals, and large families. As a result, adjusting the eligibility criteria for food assistance programs, reduced conditionality to ensure that emergency aid reached a broader population segment. However, these crisis-driven adaptations also highlighted disparities in treatment among different immigrant groups, raising concerns about biases in food aid distribution and the need for more inclusive, needs-based approaches. Municipalities have played a critical role in redefining food security responses at the local level, moving beyond traditional food banks and charity-based models to community-driven and participatory approaches. Municipal SLBs are increasingly focusing on coordinating public social services and third-sector food aid, integrating food education, initiatives. A *food security portfolio* has emerged from the crises, including municipal or third-sectors food vouchers and bags, public and private solidarity markets, school educational programs, and volunteer community kitchens, aiming to empower people facing food vulnerability rather than treating them solely as aid recipients. These initiatives reflect a shift toward participatory governance, where local governments collaborate with civil society organizations, businesses, and food system stakeholders to co-create sustainable solutions. However, challenges remain in terms of fully institutionalizing these changes. Many local initiatives operate outside formal welfare structures, relying on short-term public funding and volunteer networks, making long-term sustainability a key concern. Additionally, while some cities have embraced a rights-based approach to food security, others still rely on traditional, top-down welfare mechanisms, limiting community engagement and collaborative learning opportunities. The transition from food poverty to food vulnerability is a crucial step toward more inclusive and sustainable food security policies at all levels of governance. However, the effectiveness of this shift depends on the effort to increasingly acknowledge food as a human right, requiring more substantial institutional support, cross-sector collaboration, and the scaling-up of successful grassroots innovations.

Enhancing SLB's social learning through a cross-sectoral governance platform and coordination: The general condition of poly-crisis has highlighted the importance of procedural innovations regarding social learning. SLB's social learning entails a process of collective and communicative learning in which social interaction results in gains in knowledge, acquisition, translation and diffusion, new skills, and interpersonal relations

(Muro & Jeffrey, 2008). The multilevel and intersectoral collaboration triggered by the pandemic created an opportunity to develop new operational approaches beyond the immediate crisis. At the European level, a multilevel platform has facilitated horizontal and vertical communication and collaboration among SLBs and stakeholders, sharing and spreading knowledge about the local implementation processes. The platform method, easing best practice exchange, also contributed to boosting social workers' skills and awareness in building greater cooperation between the public sector, the third sector, and FAP final beneficiaries. In some local contexts, this led primarily to substantive improvements—such as introducing new housing support or family assistance—without significant changes to how institutions interacted or collaborated. In contrast, other urban areas used the crisis to enhance substantive and procedural dimensions. These contexts expanded the range of services, restructured collaboration methods, strengthened coordination among actors, and deepened mutual trust. This procedural evolution has reinforced local governance capacity and improved long-term resilience. The pandemic underscored the importance of public institutions as coordinators in food security governance, fostering more integrated and systematic interventions. Despite the advancements, the public role at the urban level has sometimes translated into active multi-agency coordination; more often, it has been a form of parallel activism to the assistance provided by the third sector.

Leveraging Third-Sector Expertise and Collaboration to Improve Trust and Social Resilience: The widespread presence of third-sector organizations in urban communities and their hands-on experience in identifying needs and reaching marginalized groups has often allowed them to operate with greater agility and contextual awareness than public institutions. In the face of recent crises, this capacity has prompted many street-level bureaucrats (SLBs) to engage more actively with third-sector and charitable organizations. Thanks to their bottom-up perspective and ability to respond swiftly, these organizations have played a key role in complementing public efforts, translating knowledge into actionable programs, and strengthening the overall response capacity. Their proximity to vulnerable populations and nuanced understanding of community dynamics have encouraged public institutions to move away from traditional top-down approaches and toward more adaptive, people-centered food security strategies.

However, this collaboration has taken two distinct forms. In some cases, a truly integrated model emerged in which public and third-sector actors coordinated operational efforts to jointly address the needs of vulnerable populations. In others, collaboration remained fragmented—public and private actors operated in parallel, engaging primarily through bureaucratic or financial channels rather than through shared planning or implementation. In these cases, the third sector played a critical but separate role, focusing on empowerment and service delivery without complete alignment with public strategies. This divergence reflects broader tensions within the trust-building process. While third-sector organizations often step in to fill critical gaps in food aid and social services, their involvement can unintentionally obscure systemic weaknesses in public welfare infrastructure. European policies acknowledge that public institutions sometimes rely heavily on these actors without offering sufficient structural support or integrating their contributions into long-term strategies. Anti-crisis measures aim to address this imbalance, but third-sector actors are still often expected to meet complex social demands with limited resources and uncertain mandates—blurring the line between supplementing and substituting state responsibilities.

Furthermore, differing organizational cultures—such as the flexible, needs-driven ethos of third-sector groups versus public agencies' bureaucratic, rules-based logic—can lead to friction in coordination, accountability, and decision-making processes. Despite these challenges, the interaction between public and third-sector actors remains an essential, albeit sometimes uneasy, driver of innovation and responsiveness in local food security governance.

Real-time service monitoring: The pandemic underscored the critical need for street-level real-time monitoring of security interventions to ensure that responses remain updated, agile, and attuned to rapidly changing conditions. As successive crises—whether economic, health-related, or environmental—vary in nature and intensity, real-time monitoring emerges as a vital procedural innovation. It enables public institutions to identify and track key variables influencing food insecurity risks, thereby supporting timely and targeted interventions. However, the implementation of such monitoring systems faces several challenges. Many local governments still rely on outdated data collection methods, fragmented information systems, and reactive decision-making processes. Institutional silos, limited interoperability between sectors, and a lack of digital infrastructure further hinder the ability to capture and analyze real-time data.

Moreover, traditional indicators—such as income thresholds or relative poverty—fail to fully capture the complexity of food insecurity in today's context. After recognizing these limitations, new European frameworks call for a broader set of metrics that reflect the multidimensional nature of food insecurity. These include age, geographical isolation, energy and educational poverty, digital exclusion, and other structural factors that influence a household's ability to access nutritious and sustainable food consistently. In this context, artificial intelligence (AI) holds significant potential to transform real-time monitoring. AI technologies can process vast and diverse datasets from multiple sources—including social services to isolated people, health records, school canteen attendance, social supermarket attendance, as well as social media—to detect patterns, forecast risks, and trigger early interventions. Nonetheless, the adoption of AI also raises concerns regarding data privacy, ethical use, and the risk of reinforcing existing biases if models are not carefully designed.

Policy Reuses and Institutionalization of Food Security Programs: Governments at multiple levels—European, national, and local—are moving toward formalizing food security measures within public policy frameworks. Rather than treating food aid as an emergency response dependent on charitable donations and surplus redistribution, food security is increasingly integrated into social protection systems, public procurement policies, and urban planning strategies. The EU's Farm to Fork Strategy and commitments to sustainable food systems encourage a shift toward state-supported food policies that prioritize resilience, sustainability, and equity over ad hoc interventions. National governments are experimenting with comprehensive food security policies, adopting food income, expanding access to more nutritious school meal programs, supporting the spread of social supermarkets, and food distribution networks with short supply chains. In some cases, countries have enhanced regulatory mechanisms to ensure that surplus food redistribution is systematically managed rather than left to voluntary initiatives. Cities are increasingly engaging in food governance through municipal food policies, food councils, and collaborative urban food strategies that integrate food security within broader welfare and sustainability agendas.

Governance Under Pressure: Learning, Trust, and Adaptation in Food Assistance Systems: In analyzing the governance dynamics of Food Assistance Programs at European, national, and urban levels through the lens of the collaborative learning framework, a common *fil rouge* emerged: the growing role played by public governance and street-level organizations as both platforms and cornerstones for multilevel collaborative governance.

The primary focus of this work has been understanding how street-level mechanisms of collaborative governance foster learning, trust-building, and adaptive capacity to meet the needs of vulnerable populations during crises. Findings highlight how under pressure, specific actions—such as rapid agenda modification, and relay on the adoption of new procedural tools for stakeholder coordination—facilitate aid delivery, monitoring, and program institutionalization, enabling public institutions to learn from non-experts, particularly third-sector organizations, and food assistance beneficiaries.

One may observe that despite advancements in discourse and governance dynamics at European, national, and urban levels, the responsibility for ensuring food access—a fundamental human right—has yet to be fully institutionalized within welfare systems, nor has it been fully integrated as a core dimension of broader food policies. While multiple crises have accelerated the salience of the issue and the adoption of cross-sectoral and collaborative practices to enhance food security governance, introducing various procedural tools and enhancing public–private coordination and cooperation, the *right to food* is still far from fully recognized or institutionalized within an integrated approach.

Under the pressure due to crises, food security has become an integral part of the broader social security framework at the European level, underscoring its role as a critical component of societal well-being, but such a transition has its strengths and weaknesses and can still be deemed incomplete. Procedurally, the crises accelerated the adoption of multilevel collaborative tools to facilitate coordination among governmental, non-governmental, and private sectors to turn food assistance from a short-term relief effort into a structured component of multilevel social governance. Over time, FEAD has evolved into a platform that not only coordinates diverse policy instruments across European, national, and local levels but also promotes social and institutional learning, cross-sector collaboration, and greater resilience within welfare systems. By integrating food aid with social inclusion measures, FEAD encourages a more holistic approach to addressing material deprivation and social vulnerability. However, several barriers continue to hinder the full institutionalization and strategic integration of food aid programs. One persistent challenge is the siloed nature of public administration, where responsibilities and data remain fragmented across departments, thus limiting coordinated action.

Additionally, awareness of food security as a key policy issue—intersecting health, social protection, and sustainability—is still limited in some institutional settings. As a result, food aid is often treated as a peripheral or temporary intervention rather than a core component of social policy. Another major issue is the continued over-reliance on third-sector organizations to deliver food assistance, often without adequate support or long-term integration into public welfare strategies. While these actors provide valuable local knowledge and agility, their role is frequently framed as a substitute rather than a complement to public responsibility, which can undermine the stability and scalability of interventions. In this perspective, crises highlighted the urgent need to rethink and revise conditionality in European resource allocation, prompting a shift toward more strategic and integrated approaches, fostering long-term resilience, embedding sustainability criteria, and enhancing social safety nets to protect vulnerable populations. Similar processes unfolded at the national and urban levels, aligning local food delivery systems with the overarching social security goals and a network of policies, programs, and services designed to protect residents from economic insecurity, social vulnerabilities, and unforeseen crises. The introduction of new conditionalities in food aid delivery was intended to ensure that food security efforts met short-term relief, in some cases driving the development of more robust, adaptive systems capable of enduring and thriving amid future crises.

During poly-crisis, multidisciplinary working groups, task forces, and joint initiatives became the norm, ensuring that food security policies were not developed in isolation but rather in tandem with broader crisis management efforts. A key feature of the revised governance model was the cross-sectoral collaboration between different policy areas, such as health, social protection, and agriculture. The interconnected nature of food insecurity with issues such as public health, labor markets, humanitarian protection, and social inequality required a more coordinated response across policy silos.

The EU's FEAD initiative has been crucial in promoting multilevel and cross-sectoral collaboration and social learning, establishing a unified framework for food security that enables member states to align policies, coordinate resources, and adopt innovative approaches. This initiative fostered top-down learning among EU institutions and upward feedback from local implementations, creating a dynamic governance system with learning flows across all levels. By connecting different governmental levels, the European FEAD has shown its capacity to adapt to crises while fostering a culture of mutual learning and trust-building. Vertical assistance and horizontal collaboration efforts were prompted by ensuring that policy learning achieved in specific contexts could be understood and possibly elicited in different contexts.

The Italian national food aid programs during the pandemic revealed that the FEAD framework and partnerships were a cornerstone for implementing new aid programs and influencing local agendas. Street-level bureaucrats (SLBs) played a key role in adapting national strategies to local contexts and facilitating bottom-up learning through collaboration with third-sector organizations. This feedback loop between the national and municipal levels, mediated by the third sector and the direct relationship with final beneficiaries, improved the adaptability and effectiveness of the national program.

At the urban level, Milan and Warsaw governance adaptations during poly-crises show a similar capacity to engage with beneficiaries. In both the cases analyzed, FEAD's impact was evident in scaling local food aid programs and fostering trust and learning at the street level. Nevertheless, the study emphasized that the collaborative learning process varies significantly, depending on how decision-makers frame food security within municipal governance and on governance architecture, integrating food security into broader policy frameworks to foster cohesive collaboration and innovative solutions. Conversely, when food security is narrowly framed as a welfare issue, it risks being marginalized, stifling innovation, and limiting the impact of policies. Warsaw, governed by a centralized, state-led welfare model, relied on bureaucratic mechanisms for resource mobilization. While efficient, these structures lacked the flexibility to foster cross-sectoral and public–private collaboration or innovation, leading to siloed and reactive responses. While frequent and collaborative interactions influence decision-making and the building of robust networks, they often elicit bureaucratic systems' biases, resistance to change, siloed communication, and a tendency to prioritize hierarchical control over adaptive and inclusive approaches.

Italy's decentralized food aid governance architecture enabled Milan to build on existing networks involving agencies and municipalities of metropolitan areas, NGOs, and grassroots organizations. This collaborative, multi-actor approach fostered innovative solutions, thus enhancing public–private–people partnerships and trust, widening the scope of interventions. Despite using similar tools, Milan's inclusive approach resulted in new modes and venues of collaboration and, therefore, stronger long-term institutionalization of knowledge acquisition, transformation, and spreading practices than Warsaw's more top-down framework.

In poly-crises, frequent public–private–people partnerships (4Ps) and communication were crucial to mobilize trust resources (mutual recognition and shared goals), expertise, knowledge, and innovation. Charitable, volunteer organizations, and the nonprofit sector provided valuable insights into vulnerable communities, while for-profit entities contributed through investments, technological advancements, and logistical capabilities. Nonprofits, in turn, offered on-the-ground insights, advocacy, and community engagement and communication. These partners needed to be coordinated by public actors and street-level bureaucrats (SLBs) to bridge capacity gaps and develop more inclusive, effective food security strategies. Active and continuous monitoring of end beneficiaries—those directly affected by food insecurity—within the management of food systems, in some cases, ensured that policies remain responsive to rapidly evolving needs. Food security programs become more adaptive and culturally relevant by incorporating feedback from communities, frontline workers, and local organizations. The bottom-up engagement enhanced the legitimacy and effectiveness of governance mechanisms, ensuring they aligned more closely with the direct experiences of vulnerable populations.

Crises and uncertainty become opportunities for governance experimentation and for different types of learning from the street level (Riccucci, 2005). The policy context marked a critical juncture at the urban level, but it only sometimes worked as a significant factor of change in urban food security interventions. It created a window of opportunity (Ladi & Tsarouhas, 2020). Still, it did not always act as a policy punctuation (Hogan et al., 2022) that changes policy framing, policy processes' trajectories, and actors' configuration.

Public organizations may operate at the limit of their capacity when confronted with pressing, often contradictory, and contingent goals, especially in turbulent and rapidly evolving circumstances and risks (Ansell et al., 2017). SLBs are compelled to innovate in high-pressure situations, devising swift solutions to address immediate threats. While adaptive responses are crucial during crises, public organizations often revert to their foundational structures and core functionalities once the turbulence subsides. Returning to traditional values and decision-making principles reflects a natural organizational tendency to re-establish routines after intense disruption (Ansell et al., 2017), highlighting the tension between contingent learning and the inertia of established institutional frameworks, which can limit the long-term reflexivity of the lessons learned during emergencies (Cohen, 2021; Cohen & Aviram, 2021; Zhang et al., 2021). As a result, valuable insights remained untapped, limiting the long-term impact on reshaping food assistance systems. This issue is further exacerbated by the limited scope of hierarchical governance and networks, which stifles the broader learning process from the end beneficiaries and constrains political agents from responding dynamically to shifting challenges and opportunities. Rather than fostering innovation and adaptability, this narrow approach reduces efforts to reactive, piecemeal strategies, ultimately undermining the potential for genuinely transformative governance (Nikishin et al., 2022). Like those of other researchers, the findings of this work indicate that during crises, public administrations operate at the street level as complex, composite systems shaped by the integration of evolving and often contradictory organizational principles through institutional layering (Christensen et al., 2016).

These insights suggest that the capacity for collaborative food security governance—particularly the kind rooted in trust—is not solely driven by the specific tools used to address needs, such as food stamps, digital platforms, food aids and grants, simplified procedures, or new regulations. Instead, it emerges from consistent, long-term communication practices and procedural mechanisms through which public agents enable the involvement of intermediate and end beneficiaries and foster ongoing operative dialogue among multiple agencies before, during, and after crises.

Finally, this work presents actionable recommendations for policymakers and practitioners aimed at driving sustainable changes through collaborative and credible learning in food security. Key recommendations include (a) strengthening multilevel formal-informal food aid governance platforms and enhancing both vertical and horizontal communication and collaboration by institutionalizing policy frameworks that connect and share public, private, and civil society actors' values and goals; (b) embedding collaborative knowledge exchange into organizational settings to transform short-term crisis responses into enduring, co-designed, and co-managed practices, formalizing feedback mechanisms and integrating lessons learned into policy frameworks; (c) prioritizing cross-sectoral collaborations in decision-making by aligning food security interventions with health, environmental, and economic strategies to address structural vulnerabilities holistically. Urban food policies disconnected from the social situations of food vulnerability are a critical gap; (d) empowering local governments by equipping public officials at the street level with the necessary resources and frameworks to institutionalize food security as a fundamental right and integral component of urban social policy, including collaboration with third-sector actors; (e) fostering public–private–people partnerships, leveraging the strengths of all sectors to co-create resilient systems that incorporate technological advancements, community engagement, and cultural relevance; (f) expanding real-time monitoring at the community level to develop adaptive service delivery systems, respond swiftly to emerging vulnerabilities, and ensure inclusion.

Addressing food insecurity in a poly-crisis context requires a shift from siloed approaches to collaborative, dynamic, and communicative governance models. Urban food policies will hopefully go beyond reactive emergency responses and begin to institutionalize lessons from crisis collaboration and communication, thus building resilient systems that address structural inequalities. By recognizing food as a fundamental human right, policymakers can develop transformative governance mechanisms that are both responsive and forward-looking. A cohesive, long-term strategy will enable cities to meet immediate needs while tackling food insecurity's root causes, thereby making societies better prepared for future challenges.

References

Ansell, C. K., Trondal, J., & Øgård, M. (2017). *Governance in turbulent times*. Oxford University Press. Retrieved December 31, 2023, from https://books.google.com/books?hl=it&lr=&id=gwuDDQAAQBAJ&oi=fnd&pg=PP1&dq=ansell+etal+2017+Governance+in+turbulent+times&ots=QtpPEcAEeA&sig=ob9NH46e5oH_9Sbx_j39y3V37ig.

Christensen, T., Danielsen, O. A., Laegreid, P., & Rykkja, L. H. (2016). Comparing coordination structures for crisis management in six countries. *Public Administration, 94*(2), 316–332.

Cohen, N. (2021). *Policy entrepreneurship at the street level: Understanding the effect of the individual*. Cambridge University Press.

Cohen, N., & Aviram, N. F. (2021). Street-level bureaucrats and policy entrepreneurship: When implementers challenge policy design. *Public Administration, 99*(3), 427–438.

Hogan, J., Howlett, M., & Murphy, M. (2022). Re-thinking the coronavirus pandemic as a policy punctuation: COVID-19 as a path-clearing policy accelerator. *Policy and Society, 41*(1), 40–52.

Ladi, S., & Tsarouhas, D. (2020). EU economic governance and Covid-19: Policy learning and windows of opportunity. *Journal of European Integration, 42*(8), 1041–1056.

Muro, M., & Jeffrey, P. (2008). A critical review of the theory and application of social learning in participatory natural resource management processes. *Journal of Environmental Planning and Management, 51*(3), 325–344.

Nikishin, E., Schwarzer, M., D'Oro, P., Bacon, P. L., & Courville, A. (2022). The primacy bias in deep reinforcement learning. In Proceeding of the 39th *International Conference on Machine Learning*. Proceedings of Machine Learning Research (PMLR), 162, 16828–16847.

Riccucci, N. M. (2005). Street-level bureaucrats and intrastate variation in the implementation of temporary assistance for needy families policies. *Journal of Public Administration Research and Theory, 15*(1), 89–111. https://doi.org/10.1093/JOPART/MUI005

Zhang, L., Zhao, J., & Dong, W. (2021). Street-level bureaucrats as policy entrepreneurs: Action strategies for flexible community governance in China. *Public Administration, 99*(3), 469–483.

References

Arndt, C., Davies, R., Gabriel, S., Harris, L., Makrelov, K., Robinson, S., Levy, S., Simbanegavi, W., van Seventer, D., & Anderson, L. (2020). Covid-19 lockdowns, income distribution, and food security: An analysis for South Africa. *Global Food Security, 26*, 100410.

Béland, D., & Schlager, E. (2019). Varieties of policy feedback research: Looking backward, moving forward. *Policy Studies Journal, 47*(2), 184–205.

Benington, J., & Moore, M. H. (2011). Public value in complex and changing times. In J. Benington & M. H. Moore (Eds.), *Public value* (pp. 1–30). Macmillan Education UK. https://doi.org/10.1007/978-0-230-36431-8_1

Bennett, L. A. (1978). Evaluation, feedback, and policy. *Journal of the American Criminal Justice Association, 41*, 7–18.

Blomkamp, E. (2018). The promise of co-design for public policy 1. In *Routledge handbook of policy design* (pp. 59–73). Routledge. Retrieved December 2, 2024, from https://www.taylorfrancis.com/chapters/edit/10.4324/9781351252928-4/promise-co-design-public-policy-1-emma-blomkamp.

Busetti, S., & Dente, B. (2017). Using process tracing to evaluate unique events: The case of EXPO Milano 2015. *Evaluation, 23*(3), 256–273. https://doi.org/10.1177/1356389017716738

Cristofoli, D., Douglas, S., Torfing, J., & Trivellato, B. (2022). Having it all: Can collaborative governance be both legitimate and accountable? *Public Management Review, 24*(5), 704–728. https://doi.org/10.1080/14719037.2021.1960736

M. S. Righettini, *The Governance of Food Security in Europe*, https://doi.org/10.1007/978-3-031-98083-1

Cristofoli, D., Meneguzzo, M., & Riccucci, N. (2017). Collaborative administration: The management of successful networks. *Public Management Review, 19*(3), 275–283. https://doi.org/10.1080/14719037.2016.1209236

Daugbjerg, C., & Kay, A. (2019). Policy feedback and pathways: When change leads to endurance and continuity to change. *Policy Sciences, 53*, 1–16.

Dunlop, C. A., Radaelli, C. M., Tosun, J., & Graziano, P. (2023). Policy learning in the European Union. Encyclopedia of European Union public policy.

Edmondson, D. L., Rogge, K. S., & Kern, F. (2020). Zero carbon homes in the UK? Analysing the co-evolution of policy mix and socio-technical system. *Environmental Innovation and Societal Transitions, 35*, 135–161. https://doi.org/10.1016/j.eist.2020.02.005

Gerlak, A. K., Heikkila, T., & Newig, J. (2020). Learning in environmental governance: Opportunities for translating theory to practice. *Journal of Environmental Policy & Planning, 22*, 1–14.

Gofen, A., & Gassner, D. (2022). Delegating power? Performance management from a process perspective. *Governance, 35*(2), 365–384.

Hirvonen, K., De Brauw, A., & Abate, G. T. (2021). Food consumption and food security during the COVID-19 pandemic in Addis Ababa. *American Journal of Agricultural Economics., 103*(3), 772–789.

Kamkhaji, J. C. (2022). How contingent learning saved EMU in 2010. In *Policy learning and the Euro* (pp. 69–97). Springer International Publishing. https://doi.org/10.1007/978-3-031-04264-5_4

Kessl, F., Lorenz, S., & Schoneville, H. (2020). Social exclusion and food assistance in Germany. In *The rise of food charity in Europe* (pp. 49–78). Policy Press.

Konstantinidis, C. (2022). Food insecurity, austerity, and household food production in Greece, 2009–2014. *Review of Social Economy, 82*(4), 497–535. https://doi.org/10.1080/00346764.2021.2025414

Lizzi R., & Righettini, M. S. (2025). Reframing e riconoscimento del diritto al cibo nelle politiche alimentari: una sfida per l'Italia tra policy legacy e nuove sperimentazioni. Ed. Einaudi, Biblioteca delle Libertà. https://www.centroeinaudi.it/

Loopstra, R. (2020). *Vulnerability to food insecurity since the COVID-19 lockdown.* The Food Foundation.

OECD. (2017). *Trust and public policy: How better governance can help rebuild public trust* (OECD public governance reviews). OECD Publishing. https://doi.org/10.1787/9789264268920-en

Pan, E. (2020). Reimagining the climate migration paradigm. *Environmental Law, 50*(4), 1173–1225. Retrieved October 27, 2024, from https://www.jstor.org/stable/27010196

Prats, M., Smid, S., & Ferrin, M. (2024). Lack of trust in institutions and political engagement: An analysis based on the 2021 OECD trust survey. *OECD*

Working Papers on Public Governance, No. 75. OECD Publishing. https://doi.org/10.1787/83351a47-en.

Schachter, H. L., Daniel, J., & Liu, R. (2017). Win-win agreements and public private infrastructure partnerships: Managerial and governance concerns. *Public Administration Quarterly, 41*(4), 643–669.

Simmet, A., & Stroebele-Benschop, N. (2023). Food Bank operations during the COVID-19 pandemic in Germany. *Journal of Hunger & Environmental Nutrition, 18*(3), 356–371.

Thomann, E. (2019). Qualitative comparative analysis (QCA) as a tool for street-level bureaucracy research. In *Research handbook on street-level bureaucracy* (pp. 370–391). Edward Elgar Publishing.

Index[1]

A

Acquisition of information, 22
Acquisition of knowledge, 17, 68, 114, 170, 178, 180, 187
Action Aid, 32, 60, 97n3, 103, 105, 108
Adaptive decision-making, 22
Adaptive governance, 14, 15, 144
At Risk of Social Exclusion Rate (AROPE), 32, 65, 66, 70, 73, 96
Austria, 76, 84, 87, 94

B

Backward mapping, 15, 33–35, 108
Behavioral adaptation, 17
Belgium, 72, 76, 80, 84, 91
Behavioral change technique in food security, 25
Beneficiary identification, 108, 110, 167
Beneficiary profiling, 116
Bottom-up feedback, 22
Bulgaria, 32, 57, 78, 79, 84
Bureaucratic discretion, 13, 29

C

Candel, Jeroun, 2, 4, 7, 10–11, 133, 161
Caritas, 68, 89, 105, 106, 122, 124, 145, 151–153, 156, 158, 170
Centralized governance, 9, 170
Civil society engagement, 91, 155, 163
Co-creation, 11, 14, 20, 169, 170
Cognitive asymmetries, 109
Collaborative decision-making, vii, 177
Collaborative governance and learning processes, 17
Collaborative governance (CG), viii, 2, 7–11, 30, 107, 162, 177, 184
Collaborative governance during crises, 9

[1] Note: Page numbers followed by 'n' refer to notes.

M. S. Righettini, *The Governance of Food Security in Europe*, https://doi.org/10.1007/978-3-031-98083-1

Collaborative learning, x–xii, 1–39, 55, 90, 117, 127, 177, 178, 180, 183, 186
Collaborative learning in food security governance, x, xii, 22, 24, 178
Collaborative learning *vs.* collective learning, 20
Common Agricultural Policy (CAP), 59, 61, 62, 178
Comparative case studies, 39
Concatenation of mechanisms (CM), 25, 26, 107
Consensus-building, 10
Contingent learning, 17, 21, 36, 188
Coronavirus Response Investment Initiatives Plus (CRII+), 70, 74, 97n4
COVID-19, v, vi, x, 1, 13, 15, 57, 59, 60, 70–72, 77–79, 88, 91, 103, 105, 106, 110, 112–114, 116, 125, 128, 133, 136–138, 144–146, 149, 150, 152, 155–158, 161–163, 165, 179
Crisis governance, 15
Crisis management, 14, 24, 31, 33, 123, 138, 142, 147, 154, 163, 166, 185
Crisis response, 11–16, 36, 106–110, 137, 142, 153–155, 157, 167, 171, 188
Cross-boundaries dynamic, 9
Cross-sector collaboration, 84, 93, 169–170, 180, 184, 185
Cyprus, 32, 80, 84, 87, 96
Czechia, 76, 84, 94, 96

D
Data integration, 168
Decentralization, 93, 143, 144
Decentralized governance, xi, 10, 90, 110
Denmark, 68, 84
Diffusion of knowledge, 27, 180
Digital divide, 124
Digital innovation, 158
Digital solutions, 120
Digital voucher, 120–121
Discretionary decisions, 12
Discretionary power, 15
Discretion in delivering public services, 8, 12
Discretion in policy implementation, x
Discretion in public service delivery, x
Discretion in responding to beneficiaries, 13
Dissemination of information, 22, 23
Double-loop learning, 18

E
Emergency food aid, 141, 183
Emergency food distribution, 36
Emergency response, 20, 37, 55, 141, 147, 150, 167, 183, 189
End beneficiaries, 11, 18, 21, 24, 32, 69, 85, 95, 107, 109, 124, 155–157, 187, 188
Equitable distribution of FEAD Resources, 95
Estonia, 32, 84
European Child Guarantee program, 71
European Federation of Food Banks (FEBA), 67
European Social Fund+ (ESF+), 63–65, 67, 70, 89, 96, 137

Evidence-based resource distribution, 65
External crises, 26

F
FAOSTAT, 57
Feedback loops, 17, 23, 35, 135, 186
Finland, 60, 72, 80, 84, 90, 94
Focusing attention, 27
Food access, 4, 60, 104, 128, 136, 141, 146, 147, 184
Food aid beneficiaries, 117
Food aid governance, 25, 61–67, 96, 136, 177, 186, 188
Food Aid Programs, x–xii, 8, 14, 20, 22, 23, 25, 31, 33, 36, 38, 58, 61, 80, 85, 87, 92, 103–129, 136, 140, 145, 166, 170, 184, 186
Food and Agriculture Organization (FAO), v, vi, 1–3, 5, 56–58, 60, 134
Food assistance programs (FAPs), vi, 25, 37, 57, 94, 96, 110, 114, 117, 119, 120, 122, 123, 147, 155, 159, 180, 183
Food availability, 4, 7, 55, 58, 60, 145
Food Bank, 67, 68, 75–77, 89–91, 94, 105, 106, 122, 141, 142, 145, 152–154, 156, 161, 166, 170, 180
Food deserts, 3, 19
Food distribution, v, 35–37, 62, 75, 78, 79, 89, 91, 107, 112, 114, 126, 127, 136, 157, 159, 161, 164, 165, 168, 169, 183
Food Income Initiative (FII), 126–128
Food insecurity, vi–ix, xi, 2–5, 19, 32, 33, 35, 55–96, 103, 106, 107, 111, 113, 119, 124, 125, 127, 133, 135–138, 140–147, 165–169, 178, 179, 182, 183, 185, 187, 189
Food Insecurity Experience Scale (FIES), 57, 60
Food poverty, 3, 32, 58, 61, 63, 67, 70, 73, 74, 76, 80, 91, 96, 103, 106, 127, 133, 139, 161, 162, 166, 168, 179, 180
Food quality, 4, 167, 179
Food security, v–xi, 1–11, 13–15, 19, 23, 25, 27, 28, 31, 32, 34–38, 55, 57–61, 65–67, 74–78, 88, 91–95, 103, 105, 107, 110–113, 115, 123, 127–129, 133–171, 177–181, 183–189
Food security governance, viii–x, xii, 1–7, 10, 22, 24, 25, 35, 38, 39, 57, 58, 60, 63–65, 74–78, 105, 112, 135–144, 147, 163–164, 168, 171, 177–179, 181, 182, 184
Food stability, 3, 4, 7, 27, 55, 58, 59, 88
Food utilization, 4
Food voucher, 89, 107, 110–112, 115, 117, 118, 120–123, 125, 140, 150, 180
Food voucher systems, 37
Food vulnerability, x, 56–61, 63, 79, 84, 96, 108, 110, 111, 154, 179, 180, 188
Formal and informal feedback, 22
Forward-looking learning, 18
France, 67, 68, 72, 73, 77, 79, 84, 96, 140–143
Frequent interaction, 28, 31, 69, 140, 164

Frontline discretion, 13
Frontline public servants, 29
Frontline workers, vii, ix, 34, 115, 121, 187
Fund for European Aid to the Most Deprived (FEAD), x, 55, 62–65, 67–96, 97n2, 104, 105, 109, 112, 113, 117, 126, 127, 137, 141–143, 145, 148, 153, 156, 161, 164, 167, 169, 170, 184–186

G
Germany, 57, 68, 72, 73, 75, 78–80, 82, 87, 90, 91, 93, 94, 96, 140–142
Global Food Security Index (GFSI), 60
Glocal, 2
Governance adaptation, 186
Governance agenda, 25
Governance complexity, ix, 10, 14, 57
Governance structures, vi, xi, 9, 10, 15, 25, 28, 63, 82, 93, 110, 142, 144, 170
Greece, 72, 79, 84, 96, 140, 141

H
Human right to food, 3, 5, 180, 184
Hungary, 76, 84, 87, 91, 94, 96
Hybrid governance models, 143

I
Implementation of food security, ix, 115, 127
Inclusive governance, 163
Inertia, 17, 188
Information acquisition, 22
Information translation, 22, 23
Institutional learning, 2, 139, 154, 169, 171, 184
Institutional responses and community needs, 164
Institutional settings, 9, 11, 15, 169, 185
Integrated Food Security Phase Classification (IPC), 60
Interactive dialogue and trust, 24
Interactive learning, 18
Inter-crisis learning, 23
Intermediate beneficiaries, 63, 118
Ireland, 57, 60, 84, 94, 96
Italy, xi, 32, 57, 68, 72, 73, 80, 82, 90, 93, 103–129, 140, 143, 144, 186

J
Joint responsibility, 10

K
Knowledge acquisition, 17, 114, 170, 178, 187
Knowledge diffusion, 27
Knowledge management, 114–115
Knowledge translation, 116

L
Latvia, 32, 79, 84
Learning by doing, 17
Learning from beneficiaries, 96
Learning triggers, 17
Legislative measures, 27
Lessons learned, 36, 124, 127, 151, 153, 165, 168, 169, 188
Lithuania, 32, 79, 84, 87, 90, 91
Local discretion, 95

M
Malta, 32, 84, 94
Mechanisms of action (MoA) in governance, 25
Mechanisms of collaborative learning, xi, xii, 24–39
Mechanistic approach, x, 24
Milan, xi, 39, 105, 143, 144, 159, 161–171, 186
Milan Urban Food Policy Pact (MUFPP), 133–135, 161, 162
Monitoring and evaluation, 34, 35, 157–158
Monitoring and evaluation tools
  AROPE, 32, 65, 66, 70, 73, 96
  FIES, 57, 60
  GFSI, 60
Multi-actor collaboration, 59
Multi-agency collaboration, 15, 108, 134
Multilevel governance, vi–ix, 61, 66, 67, 112, 138, 141, 145, 150, 178
Multi-stakeholder participation, 2

N
Netherlands, 68, 84, 87, 91, 136
Non-expert, 18–24, 184
Non-governmental organizations (NGOs), vii, 20, 24, 28, 55, 60, 67–69, 85, 92, 109, 115, 143, 147–149, 152–156, 158–160, 164, 186
Nutritional Standards, 68
Nutritional status, 3, 5

O
Operational programme (OP), 62, 63, 67, 71, 72, 74, 127
Organizational collaborations, 29
Organizational learning, 18
Organizational networking, 78

P
Pandemic, v–vii, ix–xi, 1, 4, 8, 13–15, 25, 27, 29, 36, 55–57, 59, 60, 63, 70, 71, 75, 77–79, 87–93, 103–129, 133–171, 177, 179, 181, 182, 186
Poland, 32, 68, 72, 73, 76–80, 84, 87, 91, 94, 96, 140, 141, 143–151, 153, 155, 156, 158, 160
Policy adaptation, ix, 7, 12, 15, 16, 20, 24, 29, 33, 35–37, 135, 139, 147, 163, 179
Policy arenas, 9, 11
Policy design and delivery, 2, 10, 34
Policy implementation, vi, x, 9, 12, 13, 34, 60, 144
Policy implementation, vi, x, 9, 12, 13, 33, 60, 144
Policy innovation, 95, 143, 170, 171
Policy learning, 16–19, 25, 94, 186
Policy networks, 16, 69
Policy reuse, 36, 167–169
Poly-crisis, vi, ix–xi, 14, 23, 38, 55–96, 138, 144–150, 177, 178, 180, 185, 189
Portugal, 32, 80, 84, 91, 94, 96
Procedural adaptability, 150
Procedural flexibility, 110
Procedural tool redesign, 15
Procedural tools, 7, 11, 15, 22, 28–30, 80, 109, 136, 165, 178, 184
Program reuse, 36
Programme for distributing Food to the Most Deprived (PEAD), 61, 62, 64–65, 104

Public policy, v, 2, 8, 12, 18, 20, 68, 177, 179, 183
Public-private partnerships (PPPs), 55, 69, 94, 104, 113, 123, 127, 142
Public service delivery, x, 11, 14
Public value, 11, 164
Public value co-creation, 11
Public welfare systems, 168

R
Real-time learning, 15, 22
Real-time monitoring, 36, 158, 182, 183, 189
Red Cross (RC), 68, 89, 105, 122–124, 166
Reflexivity and resilience, 15
Reflexivity and responsiveness, 15
Reflexivity in governance, 15
Resilient food system, vi, 141
Return to routine, 37, 124–125
Routine change, 16, 17
Russia, 57, 71, 144, 160

S
Slovakia, 32, 84, 87, 96
Social accountability, 13
Social inclusion, 61–63, 67–69, 74, 80, 84, 85, 88, 93, 94, 96, 104, 125, 178, 179, 184
Social innovation, 63
Social services, xi, 69, 70, 79, 81, 94, 109, 111, 112, 119, 121–124, 126, 148, 155, 157, 180, 182, 183
Social vulnerability, 55, 68, 75, 92, 95, 96, 179, 184, 185
Spain, 72, 73, 80, 84, 87, 91, 93, 140
Stakeholder-based decision-making, 8, 134
Stakeholder certification, 28
Stakeholder interaction and certification, ix, 10, 28, 33, 109, 140
Stakeholders willingness to share, 11
Street level bureaucrats (SLBs), vii–xi, 2, 7, 11–21, 23, 24, 28, 29, 31–34, 38, 39n1, 68, 74, 78–82, 84, 89, 90, 93–95, 106–108, 110, 113, 115, 117, 118, 122, 124, 128, 140, 141, 155, 158, 177, 178, 180, 181, 186, 187
Street-level collaborative learning, x, 1–39
Street-level food aid governance tools, 96, 136
Street-level innovation, 94, 171
Street-level public servants, 7
Street-level service delivery, 11, 189
Substantive tools, 165
Sustainable food aid, 68
Sustainable food governance, 136
Sweden, 68, 84, 96

T
Target groups
  children, 85
  homeless, 85
  migrants, 85
Targeting and screening, 14, 109, 113, 117–119
Target recognition, 31, 108, 155
Theoretical and methodological approaches, 4
Third sector, 1, 14, 55, 67, 89–91, 94, 105, 111–113, 118, 122–127, 145, 161, 166–168, 170, 179–182, 184–186, 189
Translating information, 23
Translation of knowledge, 114, 178
Trust-building strategies, 29
Trust-building through collaborative governance, 93

Trust dynamics, 30
Trust in food aid delivery, 20
Trust in law enforcement SLBs, 20
Trust mobilization, 28

U
Ukraine, 57, 71, 136, 144, 147, 148, 150
United Nations High Commissioner for Refugees (UNHCR), 147, 160
Urban food aid programs, 34
Urban food insecurity, 137
Urban governance, 2, 105, 136, 138, 139, 141, 143, 164
Urban resilience, 142

V
Vulnerability, v, vii, x, 8, 16, 20, 30, 32, 33, 55–61, 63, 65–68, 72, 73, 75, 77, 79, 80, 84, 87, 91, 92, 95, 96, 106, 108, 110, 111, 115, 118, 135, 136, 139, 142–144, 154–156, 161, 166, 179, 180, 184, 185, 188, 189
Vulnerable populations, vi, vii, 3, 21, 26, 33, 36, 38, 57, 59, 61, 63, 68, 71, 76, 79, 80, 89, 92, 95, 104, 111, 113, 117, 119, 120, 124, 128, 139, 145, 147, 150, 152, 158, 165–170, 178, 181, 182, 184, 185, 187

W
Warsaw, xi, 39, 143, 144, 147–161, 169–171, 171n2, 178, 186, 187
Welfare policy, 91, 93, 104, 139
Welfare systems, vii, xi, 69, 135, 136, 139, 143, 145, 150, 168, 169, 179, 184
Wicked problems, 16, 24, 31

FSC
www.fsc.org
MIX
Papier | Fördert
gute Waldnutzung
FSC® C083411

Kolibri 360 GmbH
Ferdinand-Jühlke-Straße 7
99095 Erfurt, Deutschland
produktsicherheit@kolibri360.de